Trends in Mathematics

Trends in Mathematics is a book series devoted to focused collections of articles arising from conferences, workshops or series of lectures.

Topics in a volume may concentrate on a particular area of mathematics, or may encompass a broad range of related subject matter. The purpose of this series is both progressive and archival, a context to make current developments available rapidly to the community as well as to embed them in a recognizable and accessible context.

Volumes of TIMS must be of high scientific quality. Articles without proofs, or which do not contain any significantly new results, are not appropriate. High quality survey papers, however, are welcome. Contributions must be submitted to peer review in a process that emulates the best journal procedures, and must be edited for correct use of language. As a rule, the language will be English, but selective exceptions may be made. Articles should conform to the highest standards of bibliographic reference and attribution.

The organizers or editors of each volume are expected to deliver manuscripts in a form that is essentially "ready for reproduction." It is preferable that papers be submitted in one of the various forms of TeX in order to achieve a uniform and readable appearance. Ideally, volumes should not exceed 350-400 pages in length.

Proposals to the Publisher are welcomed at either:

Mathematics Department
Birkhäuser Boston
675 Massachusetts Avenue
Cambridge, MA 02139
math@birkhauser.com

Mathematics Department
Birkhäuser Verlag AG
PO Box 133
CH-4010 Basel, Switzerland
math@birkhauser.ch

Titles in the Series

ANDERSSON/LAPIDUS (eds). Progress in Inverse Spectral Geometry

JAIN/RIZVI (eds). Advances in Ring Theory

ROSS/ANDERSON/LITVINOV/SINGH/SUNDER/WILDBERGER (eds). Harmonic Analysis and Hypergroups

Harmonic Analysis and Hypergroups

K. A. Ross
J. M. Anderson
G. L. Litvinov
A. I. Singh
V. S. Sunder
N. J. Wildberger
Editors

with the assistance of
Alan L. Schwartz
Martin E. Walter

Springer Science+Business Media, LLC

K. A. Ross
Dept. of Mathematics
University of Oregon
Eugene, OR 97403

J. M. Anderson
Dept. of Mathematics
University College London
London, UK

G. L. Litvinov
Centre for Optimization
and Mathematical Modeling
Institute for Technologies
Moscow, Russia

A. I. Singh
Dept. of Mathematics
University of Delhi
Delhi, India

V. S. Sunder
The Institute of Math. Sciences
CIT Campus, Taramani
Chennai, India

N. J. Wildberger
School of Mathematics
Univ. of New South Wales
Sidney, NSW, Australia

Library of Congress Cataloging-in-Publication Data

Harmonic Analysis and Hypergroups (International Conference on Harmonic Analysis 1995 : University of Delhi)
International Conference on Harmonic Analysis / editors, K. A. Ross . . . [et al.].
p. cm. -- (Trends in mathematics)
Proceedings of a conference held Dec. 18-22, 1995 at the University of Delhi.
Includes bibliographical references

1. Harmonic analysis--Congresses. I. Ross, Kenneth A. II. Title III. Series.
QA403.I55 1995 97-31839
515'.2433--dc21 CIP

ISBN 978-1-4899-0158-3 ISBN 978-0-8176-4348-5 (eBook)
DOI 10.1007/978-0-8176-4348-5

AMS Codes: P 43-06, 60-06, S28-06, 42-06

Printed on acid-free paper

Originally published by Birkhäuser in 1998.
Softcover reprint of the hardcover 1st edition 1998

Reformatted and typeset in LaTeX

9 8 7 6 5 4 3 2 1

Contents

Preface

Under the guidance and inspiration of Dr. Ajit Iqbal Singh, an International Conference on Harmonic Analysis took place at the University of Delhi, India, from December 18 to 22, 1995. Twenty-one distinguished mathematicians from around the world, as well as many from India, participated in this successful and stimulating conference.

An underlying theme of the conference was hypergroups, the theory of which has developed and been found useful in fields as diverse as special functions, differential equations, probability theory, representation theory, measure theory, Hopf algebras and quantum groups. Some other areas of emphasis that emerged were harmonic analysis of analytic functions, ergodic theory and wavelets.

This book includes most of the proceedings of this conference. I chaired the Editorial Board for this publication; the other members were J. M. Anderson (University College London), G. L. Litvinov (Centre for Optimization and Mathematical Modeling, Institute for New Technologies, Moscow), Mrs. A. I. Singh (University of Delhi, India), V. S. Sunder (Institute of Mathematical Sciences, C.I.T., Madras, India), and N. J. Wildberger (University of New South Wales, Australia). I appreciate all the help provided by these editors as well as the help and cooperation of our authors and referees of their papers. I especially appreciate technicial assistance and advice from Alan L. Schwartz (University of Missouri - St. Louis, USA) and Martin E. Walter (University of Colorado, USA). Finally, I thank our editor, Ann Kostant, for her help and encouragement during this project.

Very special thanks go to Dr. Ajit Iqbal Singh for organizing this conference. This was her brainchild and it would not have taken place without her determination and persistence. She was ably assisted by many colleagues, especially Dinesh Singh and B. S. Yadav, and by several cheerful and helpful students. Around 70 different people served on one or more of the ten committees responsible for this conference. For generous financial assistance, we are grateful to the (Delhi) University Grants Commission, Council of Scientific and Industrial Research, Department of Science and Technology, Indian National Science Academy, and the National Board of Higher Mathematics. We are thankful to the Department of Science and Technology for a special grant and for publishing the conference brochure.

K. A. Ross

De Branges Modules in $H^2(C^k)$

Sanjeev Agrawal and Dinesh Singh

1. Introduction

One of the most important results in invariant subspace theory is the famous "Beurling's Theorem" [1], characterizing the invariant subspaces of the shift operator S (i.e. multiplication by the coordinate function z) on the Hardy space $H^2(T)$.

Beurling's Theorem. *Let M be an invariant subspace of S in $H^2(T)$. Then there exists an inner function $q(z)$ in $H^\infty(T)$ (that is, $\{q(z)z^n\}$ is an orthonormal set) such that*

$$M = q(z)H^2(T).$$

Recently, L. de Branges [3] has generalized this theorem, generalizing, at the same time, the vector-valued version of Beurling's Theorem due to Lax and Halmos [4, 5, 7]. L. de Branges' theorem (scalar version) characterizes the class of Hilbert spaces that are contractively contained in $H^2(T)$ and on which the shift operator acts as an isometry.

De Branges' Theorem (Scalar version). *Let M be a Hilbert space contractively contained in $H^2(T)$ such that $S(M) \subset M$ and S acts as an isometry on M. Then there exists a $b(z)$ in the unit ball of $H^\infty(T)$ (unique up to a scalar factor of unit modulus) such that*

$$M = b(z)H^2(T)$$

and the norm on M is given by $\|b(z)f(z)\|_M = \|f(z)\|_{H^2}$.

In [10] we have extended the theorem of de Branges (scalar version) to the context of various Banach spaces of analytic functions on the unit disc such as the Hardy space, the Dirichlet space, the Bergman space, BMOA and VMOA. In this paper we have looked at the nature of those Hilbert spaces that are vector subspaces of the space $H^2(C^k)$ and on which S acts as an isometry. Not all such spaces will look like the space

---

[0] *Mathematics Subject Classification (1991 revision). Primary 47A15, 47B37.*

Key words and phrases. C^k-valued de Branges' Theorem, C^k-modules, shift, Hilbert C^k-modules.

as characterized in de Branges' theorem. We show precisely which spaces will look like the space as in de Branges' theorem in terms of the norm and the vector space structure. We also give examples of spaces that, as sets, look like the space in de Branges' theorem but whose geometry is different from the type as given in his theorem.

It should be borne in mind that, unlike the result of de Branges, our results do not make any contractivity or continuity assumption between the space to be characterized and the space $H^2(C^k)$.

2. Preliminary results and definitions

C^K will denote the k-dimensional unitary space. For $\alpha = (\alpha_1, \ldots, \alpha_k)$ and $\beta = (\beta_1, \ldots, \beta_k) \in C^k$, if we define

$$\begin{aligned} \alpha^* &= \overline{\alpha} = (\overline{\alpha}_1, \ldots, \overline{\alpha}_k) , \\ \alpha\beta &= (\alpha_1\beta_1, \ldots, \alpha_k\beta_k) , \\ \|\alpha\|_k^2 &= |\alpha_1|^2 + \ldots + |\alpha_k|^2 , \end{aligned}$$

then $(C^k, \|.\|_k)$ forms a Banach *-algebra. The set $\{e_i\}$ will denote the standard orthonormal basis of C^k where the ith coordinate in e_i is 1 and all the other coordinates are zero. The basis is ordered by i. All matrices in this paper are written with respect to the basis $\{e_i\}$. Let A be a nonsingular $k \times k$ matrix. For $\alpha, \beta \in C^k$, we define

$$\langle \alpha, \beta \rangle_A = \langle A\alpha, A\beta \rangle_k ,$$

It is easy to check that $\langle .,. \rangle_A$ is an inner product on C^k. The corresponding norm on C^k will be denoted by$\|.\|_A$.

Lemma 2.1. *If $\langle .,. \rangle$ is an inner product on C^k, then there exists a unique upper triangular $k \times k$ matrix A whose diagonal entries are positive real numbers and such that*

$$\langle .,. \rangle = \langle .,. \rangle_A .$$

Proof. The result follows from the characterization of positive definite sesquilinear forms on C^k and the QR factorization of matrices. For the QR factorization of matrices we refer to the following theorem in [6, page112]. If A is a nonsingular $k \times k$ matrix, there exists a unitary matrix Q and an upper triangular matrix R such that $A = QR$. R may be chosen so that all its diagonal entries are positive and, in this event, the factors Q and R are both unique.

$H^2(C^k)$ will denote the space of all formal power series $\sum \alpha_n z^n$ where $\alpha_n \in C^k$ and $\sum \|\alpha_n\|_k^2 < \infty$. It is well known that $(H^2(C^k), \|.\|_{H^2}^2)$ is a Hilbert space, where $\|\sum \alpha_n z\|_{H^2}^2 = \sum \|\alpha_n\|_k^2$. Further, $H^2(C^k)$ is a C^k-module under the multiplication $\alpha f = \alpha \sum \alpha_n z^n = \sum \alpha \alpha_n z^n$.

Let S denote the operator of multiplication by the coordinate function z on $H^2(C^k)$, that is, $S(f(z)) = zf(z)$. S is an isometry in $H^2(C^k)$. $H^\infty(C^k)$ will denote the space of all $f \in H^2(C^k)$ such that $fg \in H^2(C^k)$ for all $g \in H^2(C^k)$; the product of f and g being defined as their formal Cauchy product. Let A be a nonsingular $k \times k$ matrix. For $f \in \sum \alpha_n z^n$, if we define $\|f\|_A^2 = \sum \|\alpha_n\|_A^2$, then it is easy to see that $(H^2(C^k), \|.\|_A)$ is a Hilbert space, such that $S(H^2(C^k)) \subset H^2(C^k)$ and S acts as an isometry on it. ■

Definition 2.2. (De Branges space) If $(M, \|.\|_M)$ is a Hilbert space that is a vector subspace of $H^2(C^k)$, then M is called a de Branges space if there exists a $b(z) \in H^\infty(C^k)$ such that

$$M = b(z)H^2(C^k)$$

and $\{\alpha_n z^n b(z)\}$ is an orthogonal sequence in M for every sequence $\{\alpha_n\}$ in C^k.

Definition 2.3. ([2]) A Hilbert space $(M, \|.\|)$ is said to be a Hilbert C^k-module if

(i) M is an algebraic (left) module over C^k,
(ii) $\langle \alpha x, y\rangle = \langle x, \overline{\alpha} y\rangle$ for every $\alpha \in C^k, x, y \in M$,
(iii) there exists a constant R such that

$$\|\alpha x\| \leq R \|\alpha\|_k \|x\| \quad \text{for every} \quad \alpha \in C^k \quad \text{and} \quad x \in M.$$

Remark 2.4. All C^k-modules in this paper will be C^k-submodules of $H^2(C^k)$ in the algebraic sense.

Remark 2.5.

(1) For $f \in H^2(C^k)$, we shall write $f_{(i)}$ for $e_i f$. Note that in all the coefficients of $f_{(i)}$, all coordinates, except possibly the ith coordinate, are zero.

(2) For a space M of the type $b(z)H^2(C^k)$, we will assume that $b_{(i)} \neq 0$ for each i. Otherwise M can be considered as a subspace of $H^2(C^r)$ for a suitable $r < k$. Thus there is no loss of generality.

Proposition 2.6. *If A is a nonsingular $k \times k$ matrix, then*

(a) $(H^2(C^k), \|.\|_A)$ *is a de Branges space.*

(b) If $b \in H^\infty(C^k)$, *then* $(M, \|.\|_M)$ *is a de Branges space where*

$$M = bH^2(C^k) \quad \text{and} \quad \|bf\|_M = \|f\|_A .$$

Proof.

(a) $H^2(C^k) = eH^2(C^k)$, where $e = (1, 1, \ldots, 1)$. Now

$$\begin{aligned} \langle \alpha e z^m, \beta e z^n \rangle_A &= \langle (A\alpha e)\, z^m, (A\beta e) z^n \rangle_{H^2} \\ &= 0 \quad \text{if} \quad m \neq n . \end{aligned}$$

Thus $(H^2(C^k), \|.\|_A)$ is a de Branges space.

(b) M is obviously a vector subspace of $H^2(C^k)$ and $S(M) \subset M$. By part (a), $(eH^2(C^k), \|\ \|_A)$ is a de Branges space. Thus

$$\begin{aligned} \langle \alpha b z^m, \beta b z^n \rangle_M &= \langle \alpha e b z^m, \beta e b z^n \rangle_M \\ &= \langle \alpha e z^m, \beta e z^n \rangle_A \\ &= 0 \quad \text{if} \quad m \neq n. \end{aligned}$$

■

3. The main results

We start with the converse of Proposition 2.6(b).

Theorem 3.1. *Let* $(M, \|.\|_M)$ *be a Hilbert space that is a vector subspace of* $H^2\left(C^k\right)$ *such that* $S(M) \subset M$ *and* S *acts as an isometry on* M.

(a) Then M *is a de Branges space if and only if Ker* S^* *is an algebraic* C^k*-submodule of* $H^2(C^k)$.

(b) In this case, $M = b(z)H^2(C^k)$ *(where* $\{\alpha_n z^n b\}$ *is an orthogonal sequence in* M *for every sequence* $\{\alpha_n\}$ *in* C^k *and if* $b_{(i)} \neq 0$ *for all* (i)*, and there exists a nonsingular* $k \times k$ *upper triangular matrix* A *such that*

$$\|bf\|_M = \|f\|_A$$

for all f *in* $H^2(C^k)$.

Remark 3.2. For $k = 1$, this theorem gives the scalar version of de Branges' Theorem without the contractivity condition. Later on we shall show the existence of spaces of the type $b(z)H^2(C^k)$ that are not de Branges spaces.

Proof.

(a) The proof of this part of the theorem is essentially a simpler version of the proof of the main theorem in [11]. Here we prove the most important part of the proof. The missing links can be found in [11]. Note that Ker $S^* = M \ominus S(M)$. We write N for $M \ominus S(M)$.We first assume that N is a C^k-submodule of $H^2(C^k)$. By the Halmos–Wold decomposition

$$M = N \oplus S(N) \oplus \dots .$$

Claim 1: If $b \in N$, then $\{\alpha_n b z^n\}$ is an orthogonal set in M for every sequence $\{\alpha_n\}$ in C^k. This is easy to show and we skip the justification.

Claim 2: For each f in C^k and for all $\alpha \in C^k$ there exist positive constants A_f, B_f such that $A_f \|\alpha f\|_2 \leq \|\alpha f\|_M \leq B_f \|\alpha f\|_{H^2}$. This follows from the fact that for any $f \in M$, the set $\left\{\alpha f : \alpha \in C^k\right\}$ is a finite-dimensional subspace of $H^2(C^k)$ and of M. (Thus M satisfies condition (iii) of Definition 2.3.)

Claim 3: If $f \in N$, then $fg \in M$ for all $g \in H^2(C^k)$. Let $f = \sum \alpha_n z^n$, $g = \sum \beta_n z^n$, and $g_n = \sum_{i=0}^{n} \beta_i z^i$. Then $g_n \longrightarrow g$ in $H^2(C^k)$ and hence $\{g_n\}$ is a Cauchy sequence in $H^2(C^k)$. Also,

$$\begin{aligned}
&\|g_m f - g_n f\|_M^2 \\
&\quad = \|\beta_m z^m f\|_M^2 + \left\|\beta_{m-1} z^{m-1} f\right\|_M^2 + \dots + \left\|\beta_{n+1} z^{n+1} f\right\|_M^2 \\
&\qquad\qquad \text{(by Claim 1)} \\
&\quad = \|\beta_m f\|_M^2 + \|\beta_{m-1} f\|_M^2 + \dots + \|\beta_{n+1} f\|_M^2 \\
&\qquad\qquad \text{(as } S \text{ is an isometry on } M\text{)} \\
&\quad \leq B_f^2 \left[\|\beta_m f\|_{H^2}^2 + \|\beta_{m-1} f\|_{H^2}^2 + \dots + \|\beta_{n+1} f\|_{H^2}^2\right] \\
&\quad \leq B_f^2 \|f\|_{H^2}^2 \left[\|\beta_m\|_k^2 + \|\beta_{m-1}\|_k^2 + \dots + \|\beta_{n+1}\|_k^2\right] \\
&\qquad\qquad \text{(as } \|\alpha f\|_{H^2}^2 \leq \|\alpha\|_k^2 \, \|f\|_{H^2}^2\text{)} \\
&\quad \leq B_f^2 \|f\|_{H^2}^2 \|g_m - g_n\|_{H^2}^2 .
\end{aligned}$$

Thus, $\{g_n f\}$ is a Cauchy sequence in the Hilbert space M and hence converges to, say, $h \in M$. Then it can be shown that [11, page 65] $h = fg$. This settles Claim 3.

In fact, we have shown that for $f(z) \in N$ and $\sum \alpha_n z^n \in H^2(C^k)$,

$$\sum_{i=0}^{n} \alpha_i z^i f \longrightarrow \left[\sum \alpha_n z^n\right] f \quad \text{in} \quad M.$$

Also note that we have shown that the linear span of $\{\alpha z^n b(z) : \alpha \in C^k \text{ and } n \geq 0\}$ is dense in $(M, \|.\|_M)$.

Now it can be shown that [11, pages 66–69] if f , $g \in N$ are such that $f_{(i)} \perp g_{(i)}$, then either $f_{(i)} = 0$ or $g_{(i)} = 0$. Let i be such that there exists $f \in N$ such that $f_{(i)} \neq 0$. Let $g \in N$. It is easy to see that $g_{(i)}$ is linearly dependent on $f_{(i)}$. Let $X = \{i :$ there exists $f \in N$ such that $f_{(i)} \neq 0\}$. For each $i \in X$, choose $q_i \in N$ such that $e_i q_i \neq 0$. For $i \notin X$, take $q_i = 0$. Let $b = e_1 q_1 + \ldots + e_k q_k$. Then it is shown in [11] that $N = \{\alpha b : \alpha \in C^k\}$, and then $M = b(z)H^2(C^k)$. Conversely, it is obvious that M is a C^k-submodule of $H^2(C^k)$. By [11, page 72], $N = \{\alpha b : \alpha \in C^k\}$. This implies that N is a C^k-submodule of $H^2(C^k)$. Note that the characterization obtained here is algebraic in nature. No knowledge about the norm on M is assumed.

(b) We define $\|.\|$ on $H^2(C^k)$ as $\|f\| = \|bf\|_M$. As $\|.\|_M$ is a norm on M, it is easy to check that $\|.\|$ is a norm on $H^2(C^k)$. Obviously S acts as an isometry in $(H^2(C^k), \|.\|)$. Also, $\{\alpha_n z^n e\}$ is an orthogonal sequence in M for every sequence $\{\alpha_n\}$ in C^k where $e = (1, 1, \ldots, 1)$. Thus $H^2(C^k)$ $(= eH^2(C^k), \|.\|)$ is a de Branges space. Let $f = \sum \alpha_n z^n \in H^2(C^k)$. Let $p_n e \longrightarrow fe$ in $(eH^2(C^k)$ $\|.\|)$. Thus $\|p_n e\| \longrightarrow \|fe\|$. As C^k is a finite-dimensional subspace of $(eH^2(C^k), \|.\|)$, C^k is a Hilbert space under $\|.\|$. Thus there exists a nonsingular $k \times k$ upper triangular matrix A such that $\|\alpha\| = \|\alpha\|_A$ for every $\alpha \in C^k$. Now

$$\begin{aligned}
\|p_n e\|^2 &= \left\|\sum_{i=0}^{n} e\alpha_i z^i\right\|^2 \\
&= \sum_{i=0}^{n} \left\|e\alpha_i z^i\right\|^2 \quad \text{(as } \{\alpha_i z^i e\} \text{ is an orthogonal sequence)} \\
&= \sum_{i=0}^{n} \|e\alpha_i\|^2 \quad \text{(as } S \text{ is an isometry)} \\
&= \sum_{i=0}^{n} \|e\alpha_i\|_A^2 \\
&= \|p_n e\|_A^2 \, .
\end{aligned}$$

Thus $\|p_n e\|_A \longrightarrow \|fe\|$. The proof of (a) also shows that $p_n e \longrightarrow fe = f$ in $(H^2(C^k, \|.\|_A)$. Thus,

$$\|bf\|_M = \|f\| = \|fe\| = \|fe\|_A = \|f\|_A \,.$$

■

Theorem 3.3. *Let A be an upper triangular $k \times k$ matrix whose diagonal elements are positive real numbers. Then the following are equivalent:*

(i) A is a diagonal matrix.

(ii) There exist positive real numbers $\lambda_1, \lambda_{2,} \ldots, \lambda_k$ such that

$$\|f\|_A^2 = \lambda_1^2 \left\|f_{(1)}\right\|_{H^2}^2 + \lambda_2^2 \left\|f_{(2)}\right\|_{H^2}^2 + \ldots + \lambda_k^2 \left\|f_{(k)}\right\|_{H^2}^2$$

for every $f \in H^2(C^k)$.

(iii) $(H^2(C^k), \|.\|_A)$ is a Hilbert C^k-module.

(iv) Every closed S-invariant C^k-submodule of $(H^2(C^k), \|.\|_A)$ is a de Branges space.

Proof. Let the diagonal entries of A be $\lambda_1, \lambda_{2,} \ldots, \lambda_k$.

(i)⇒(ii). As $Ae_i = \lambda_i e_i$, it is obvious that (i) ⇒(ii).

(ii)⇒(iii). It is easy to see that $(H^2(C^k), \| \;\;.\;\; \|_A)$ is a Hilbert C^k-module if and only if $\langle e_i, e_j \rangle_A = 0$ for all $i \neq j$. Now

$$\begin{aligned} 4 \langle e_i, e_j \rangle_A &= \|e_i + e_j\|_A^2 - \|e_i - e_j\|_A^2 + i \|e_i + ie_j\|_A^2 - i \|e_i - ie_j\|_A^2 \\ &= \left(\lambda_i^2 + \lambda_j^2\right) - \left(\lambda_i^2 + \lambda_j^2\right) + i \left(\lambda_i^2 + \lambda_j^2\right) - i \left(\lambda_i^2 + \lambda_j^2\right) \\ &= 0 \,. \end{aligned}$$

(iii)⇒(iv). If $b \in \text{Ker } S^*$ $(= M \ominus S(M))$, $\alpha \in C^k$, and $g \in M$, then

$$\langle \alpha b, zg \rangle_A = \langle b, \overline{\alpha} zg \rangle_A = 0.$$

Thus Ker S^* is a C^k-submodule of $H^2(C^k)$. So, by Theorem 3.1, M is a de Branges space.

(iv)⇒(i). Let the (i, j)th entry of A be a_{ij}. Suppose that A is not a diagonal matrix. Then there exist i, j $(i < j)$ such that $a_{ij} \neq 0$. Choose (iv)⇒(i). Let the (i,j)th entry of A be a_{ij}. Suppose that A is not a diagonal matrix. Then there exist i,j $(i<j)$ such that $a_{ij}\neq 0$. Choose the "first" entry a_{ij} (that is, with least j and then least i) out of the nonzero entries above the diagonal. Note that $\min(i,j)>1$. Then

$$Ae_i=\lambda_i e_i\ ,$$

and

$$Ae_j=a_{ij}e_i+a_{i+1,j}e_{i+1}+\ldots+a_{jj}e_j\ . \qquad \text{(Note that } a_{ij}=\lambda.\text{)}$$

Let

$$\begin{aligned} bz &= \frac{1}{\lambda_i}e_i+\left[\frac{-a_{ij}}{\lambda_i\lambda_j}e_i+\frac{1}{\lambda_j}e_j\right]z \\ &= \beta_1 e_i+(\beta_2 e_i+\beta_3 e_j)\,z. \end{aligned}$$

Let M be the smallest closed C^k-submodule of $\left(H^2\left(C^k\right),\|.\|_A\right)$ containing the set $\{\alpha z^n b:\alpha\in C^k \text{ and } n\geq 0\}$. Then $S(M)\subset M$. We are given that M is a de Branges space. Now if $\alpha=(\alpha_1,\ldots,\alpha_k)$ and $n\geq 1$, then

$$\begin{aligned} \langle b,\alpha z^n b\rangle_A &= \left\langle \beta_1 e_i+(\beta_2 e_i+\beta_3 e_j)\,z,\alpha_i\beta_1 e_i z^n+(\alpha_i\beta_2 e_i+\alpha_j\beta_3 e_j)\,z^{n+1}\right\rangle_A \\ &= \langle A\,(\beta_2 e_i+\beta_3 e_j)\,z, \qquad A\,(\alpha_i\beta_1 e_i)\,z^n\rangle_{H^2} \\ &= \left\langle A\,(\beta_2 e_i+\beta_3 e_j), \qquad A\,(\alpha_i\beta_1 e_i)\,z^{n-1}\right\rangle_{H^2}\ . \end{aligned}$$

This is zero if $n\neq 1$. For $n=1$, we get

$$\begin{aligned} \langle b,\alpha z^n b\rangle_A &= \langle A\,(\beta_2 e_i+\beta_3 e_j), \qquad A\,(\alpha_i\beta_1 e_i)\rangle_{H^2} \\ &= \langle \lambda_i\beta_2 e_i+\beta_3\,(a_{ij}e_i+a_{i+1,j}e_{i+1}+\ldots+a_{jj}e_j)\,,\lambda_i\alpha_i\beta_1 e_i\rangle_{H^2} \\ &= \langle \lambda_i\beta_2 e_i+\beta_3 a_{ij}e_i,\lambda_i\alpha_i\beta_1 e_i\rangle_{H^2} \\ &= \left\langle \lambda_i\frac{-a_{ij}}{\lambda_i\lambda_j}e_i+\frac{1}{\lambda_j}a_{ij}e_i,\lambda_i\alpha_i\frac{1}{\lambda_i}e_i\right\rangle_{H^2} \\ &= 0\ . \end{aligned}$$

Thus $b\in M\ominus S(M)$. Then $e_i b\in M\ominus S(M)$ by Theorem 3.1. Hence

$$\begin{aligned} 0 &= \left\langle \beta_1 e_i+\beta_2 e_i z,\beta_1 e_i z+(\beta_2 e_i+\beta_3 e_j)\,z^2\right\rangle_A \\ &= \left\langle A\,(\beta_1 e_i)+A\,(\beta_2 e_i)\,z,A\,(\beta_1 e_i)\,z+A\,(\beta_2 e_i+\beta_3 e_j)\,z^2\right\rangle_{H^2} \\ &= \langle A\,(\beta_1 e_i)\,,A\,(\beta_1 e_i)\rangle_{H^2} \\ &= \beta_2\lambda_i\beta_1\lambda_i \\ &= \frac{-a_{ij}}{\lambda_j}. \end{aligned}$$

This is a contradiction. Thus A is a diagonal matrix. ■

Remark 3.4. Parts (i) and (iv) of the theorem show that there is an abundance of spaces satisfying the hypothesis of Theorem 3.1 that are not de Branges spaces. In fact, the final part of the proof gives a space of the type $b(z)H^2(C^k)$ that is not a de Branges space.

Corollary 3.5. *(to Theorem 3.3) Let M satisfy the hypothesis of Theorem 3.1. Then the following are equivalent.*

(i) M is a Hilbert C^k-module.

(ii) A is a diagonal matrix.

(iii) There exists $c(z) \in H^\infty(C^k)$ such that $\{\alpha_n z^n c(z)\}$ is an orthogonal sequence in M for every sequence $\{\alpha_n\}$ in C^k and

$$M = c(z)\, H^2\left(C^k\right) \text{ and } \|cf\|_M = \|f\|_{H^2} \text{ for all } f \text{ in } H^2(C^k)$$

Proof. It is easy to see that M is a Hilbert C^k-module if and only if $\left\langle b_{(i)}, b_{(j)} \right\rangle_M = 0$ for all $i \neq j$. But $\left\langle b_{(i)}, b_{(j)} \right\rangle_M = \langle e_i, e_j \rangle_A$ for all i and j. Thus Theorem 3.3 gives that (i) is equivalent to (ii).

Now suppose that A is a diagonal matrix whose diagonal entries are $\lambda_1, \lambda_2, \ldots, \lambda_k$ respectively. Then $\left\| b_{(i)} \right\|_M = \lambda_i$. Let $c = \sum_{i=1}^k \frac{b_{(i)}}{\lambda_i}$. Then it is obvious that $c(z)H^2(C^k) = b(z)H^2(C^k)$. Now,

$$\begin{aligned}
\|cf\|_M^2 &= \left\| c_{(1)} f_{(1)} + c_{(2)} f_{(2)} + \ldots + c_{(k)} f_{(k)} \right\|_M^2 \\
&= \left\| \frac{1}{\lambda_1} b_{(1)} f_{(1)} + \frac{1}{\lambda_2} b_{(2)} f_{(2)} + \ldots + \frac{1}{\lambda_k} b_{(k)} f_{(k)} \right\|_M^2 \\
&= \left\| b(z) \left(\frac{1}{\lambda_1} f_{(1)} + \frac{1}{\lambda_2} f_{(2)} + \ldots + \frac{1}{\lambda_k} f_{(k)} \right) \right\|_M^2 \\
&= \left\| \frac{1}{\lambda_1} f_{(1)} + \frac{1}{\lambda_2} f_{(2)} + \ldots + \frac{1}{\lambda_k} f_{(k)} \right\|_A^2 \\
&= \left\| f_{(1)} \right\|_{H^2}^2 + \left\| f_{(2)} \right\|_{H^2}^2 + \ldots + \left\| f_{(k)} \right\|_{H^2}^2 \qquad \text{(see Theorem 3.3(ii))} \\
&= \|f\|_{H^2}.
\end{aligned}$$

Thus (i) and (ii) imply (iii).

To see that (iii) implies (i), we note that $\langle \alpha cf, cg \rangle_M = \langle \alpha f, g \rangle_{H^2} = \langle f, \overline{\alpha} g \rangle_{H^2} = \langle cf, \overline{\alpha} cg \rangle_M$ ■

Acknowledgements. The second author would like to thank the Indian Statistical Institute, Delhi Centre, for providing a stimulating mathematical atmosphere and excellent facilities while this paper was being completed. He would also like to thank the University Grants Commission for a research grant under the Career Award.

References

[1] A. Beurling, *On two problems concerning linear transformations in Hilbert space*, Acta. Math. **81** (1949), 239–255.

[2] F.F. Bonsall and J. Duncan, *Complete Normed Algebras,* Springer Verlag, 1973.

[3] L. de Branges, *Square Summable Power Series,* Springer Verlag (to appear).

[4] H. Helson, *Lectures on Invariant Subspaces.* Academic Press, 1964.

[5] K. Hoffman, *Banach Spaces of Analytic Functions*, Prentice Hall, 1962.

[6] R.A. Horn and C.R. Johnson, *Matrix Analysis,* Cambridge University Press, Cambridge, 1990.

[7] M. Rosenblum and J. Rovnyak, *Hardy Classes and Operator Theory,* Oxford University Press, 1985.

[8] D. Singh, *Brangesian Spaces in the Polydisc,* Proc. Amer. Math. Soc. **110** (1990), 971–977.

[9] D. Singh and S. Agrawal, *De Branges modules in l^2-valued Hardy spaces of the circle and the torus,* Journal of Mathematical Sciences (U.N. Singh Memorial Volume) **28** (1994), 235–266.

[10] D. Singh and S. Agrawal, *De Branges spaces contained in some Banach spaces of analytic functions,* Illinois Journal of Math. **39** (1995), 351–357.

[11] B.S. Yadav, D. Singh and S. Agrawal, *De Branges Modules in $H^2(C^k)$ of the Torus* IN *Functional Analysis and Operator Theory,* Lecture Notes in Mathematics (Number **1511**), Springer Verlag (1992), 55–74.

Dept. of Mathematics, St. Stephen's College, Delhi-110007, India

E-mail : dinesh@isid.ernet.in; Department of Mathematics, University of Delhi, Delhi-110007, India. Currently at Indian Statistical Institute, 7 S.J.S. Sansanwal Marg, New Delhi-110016, India

Some Methods to Find Moment Functions on Hypergroups

Léonard Gallardo

Abstract

We present two methods to find moment functions on hypergroups. The first consists in the determination of admissible paths in the dual and the second is based on the property of asymptotic drift of the convolution. Some illustrative examples are also considered.

1. Introduction

In classical probability theory the usefulness of the notions of expectation and variance is due to the additivity property. Consider now a hypergroup $(K, *)$ and suppose there exists a pair (m_1, m_2) of measurable real-valued and locally bounded functions on K such that

$$(1.1) \qquad \langle \delta_x * \delta_y, m_1 \rangle = m_1(x) + m_1(y)$$

$$(1.2) \qquad \langle \delta_x * \delta_y, m_2 \rangle = m_2(x) + 2m_1(x)m_1(y) + m_2(y)$$

$$(1.3) \qquad m_1^2(x) \leq m_2(x),$$

for all $x, y \in K$ (δ denotes Dirac measure). Such functions m_1 and m_2 are called associated moment functions of order 1 and 2 respectively. We will say more briefly that (m_1, m_2) is a pair of moment functions. With respect to this pair and for any probability measure $\mu \in M_1(K)$ on K with $\langle \mu, m_2 \rangle < +\infty$, we define its generalized expectation and variance by

$$(1.4) \qquad \mathbb{E}_*(\mu) = \langle \mu, m_1 \rangle$$

$$(1.5) \qquad V_*(\mu) = \langle \mu, m_2 \rangle - \langle \mu, m_1 \rangle^2.$$

For a K-valued random variable X distributed according to μ, we also denote

$$(1.6) \qquad \mathbb{E}_*(X) = \mathbb{E}_*(\mu) \text{ and } V_*(X) = V_*(\mu).$$

From properties (1.1) to (1.3) it follows immediately that the operators $\mathbb{E}_*$ and V_* have the additive property; that is, for $\mu_i \in M_1(K)$ such that $\langle \mu_i, m_2 \rangle < +\infty \quad (i = 1, 2)$,

$$\mathbb{E}_*(\mu_1 * \mu_2) = \mathbb{E}_*(\mu_1) + \mathbb{E}_*(\mu_2). \tag{1.7}$$

$$V_*(\mu_1 * \mu_2) = V_*(\mu_1) + V_*(\mu_2). \tag{1.8}$$

Such generalized expectations and variances were used in the 1960s by Karpelevich et al. to study a central limit theorem in the Poincaré–Lobachevski space [Ka] and then by many other authors ([F], [T1] etc.), but the fundamental importance of this notion for the study of probability limit theorems on hypergroups was clearly revealed by Zeuner in 1992 [Z]. We can summarize his ideas as follows:

Let (S_n) be a discrete time increment process on the hypergroup $(K, *)$; that is, for each $n \in \mathbb{N}$ and each Borel set $A \subset K$ we have

$$\mathbb{P}(S_n \in A | \mathcal{F}_{n-1}) = (\delta_{S_{n-1}} * \nu_n)(A) \text{ a.s.},$$

where $(\mathcal{F}_n)$ is the canonical filtration of (S_n) and (ν_k) is a sequence of probability measures on K. If $\nu_k = \nu$ (independent of k), (S_n) is a random walk of law ν. For such a process, if (m_1, m_2) is a pair of moment functions on $(K, *)$, the realvalued stochastic processes

$$(m_1(S_n) - \mathbb{E}_*(S_n); n \in \mathbb{N})$$

and

$$(m_2(S_n) - 2\mathbb{E}_*(S_n)m_1(S_n) - V_*(S_n); n \in \mathbb{N})$$

are martingales (i.e. processes very close to processes of sums of independent real random variables). They can then be analyzed with classical tools in order to get information on the asymptotic behavior of the trajectories of (S_n). For example, if (S_n) is a random walk on K of law ν with $V_*(\nu) < +\infty$, we have

$$\lim_{n \to \infty} \frac{m_1(S_n) - \mathbb{E}_*(S_n)}{n} = 0 \text{ a.s.}$$

([Z], Theorem 7.2, p. 400).

The problem of how to find a pair of moment functions on a hypergroup is far from trivial. This question has not yet been studied for itself. In general, moment functions are determined on each particular example by ad-hoc methods. In this paper we will review what can be said about this problem.

At present there are two known methods to construct moment functions on commutative hypergroups. The first one requires investigating the dual of the hypergroup. The other is more geometrical but works only on a class of hypergroups with a convolution having asymptotic drift at infinity [G2]. For some examples the two methods give rise to the same moment functions, but the problem of uniqueness is not yet clear. In fact, our paper does not give exhaustive answers except perhaps for the case of polynomial hypergroups where our results are probably new and the problem of the determination of moment functions is completely solved. The structure of our paper is as follows:

Section 2 contains generalities about hypergroups and some properties of moment functions, Section 3 reviews the known methods of finding moment functions and Section 4 investigates three typical examples with explicit calculations.

2. Generalities

2.1. Definitions

In this paper we consider hypergroups in the sense of Jewett [B-H]. For the sake of completeness, recall that $(K,*)$ is a hypergroup if K is a locally compact topological space with an operation $*$, called convolution, on the space $M(K)$ of bounded complex Radon measures on K. The operation $*$ is a bilinear and separately continuous mapping from $M(K)\times M(K)$ into $M(K)$ and preserves probability measures (i.e. $M_1(K) * M_1(K) \subset M_1(K)$). Moreover, the following axioms must be satisfied:

C1 $0\delta_x * (\delta_y * \delta_z) = (\delta_x * \delta_y) * \delta_z \qquad (\forall x, y, z \in K)$.

C2 There exists an element $e \in K$ such that $\delta_e * \delta_x = \delta_x * \delta_e = \delta_x$.

C3 There exists a continuous involution $x \to x^-$ of K such that $e \in \mathrm{supp}(\delta_x * \delta_y)$ if and only if $x = y^-$ and the canonical extension of this involution to $M(K)$ satisfies $(\delta_x * \delta_y)^- = \delta_{y^-} * \delta_{x^-}$.

C4 The mapping $(x,y) \to \delta_x * \delta_y$ is weakly continuous from $K \times K$ into $M_1(K)$.

C5 $\mathrm{supp}(\delta_x * \delta_y)$ is compact and the mapping $(x,y) \to \mathrm{supp}(\delta_x * \delta_y)$

from $K \times K$ into the set of nonvoid compact subsets of K (endowed with the Michael topology) is continuous.

A locally compact topological group G is a hypergroup with natural convolution $\delta_x * \delta_y = \delta_{xy}$.

$(K,*)$ is commutative if $\delta_x * \delta_y = \delta_y * \delta_x (\forall x, y \in K)$. In particular, if $x^- = x$ we say that K is Hermitian; in this case it is a commutative hypergroup.

A commutative hypergroup always has a Haar measure [S], i.e. a positive Radon measure ω on K such that

$$\langle \omega, \delta_x * f \rangle = \langle \omega, f \rangle, \tag{2.1.1}$$

for every $f \in C_c(K)$ and $x \in K$ where $\delta_x * f$ is the function defined by $(\delta_x * f)(y) = \langle \delta_{x^-} * \delta_y, f \rangle$.

For a commutative hypergroup $(K,*), \mathcal{X}(K)$ denotes the set of all continuous functions: $\chi : K \to \mathbb{C}$ such that for all $x, y \in K$

$$\langle \delta_x * \delta_y, \chi \rangle = \chi(x)\chi(y). \tag{2.1.2}$$

The dual $\hat{K}$ of K is then the set of those $\chi \in \mathcal{X}(K)$ that are bounded and satisfy $\chi(x^-) = \overline{\chi(x)}$, for all $x \in K$. Such functions χ are called characters. A Fourier analysis can then be developed, but we will not need it in this paper (see [B-H] for details).

2.2. Examples

2.2.1. Chebli-Trimèche hypergroups:

Let A be an increasing unbounded real function on $\mathbb{R}_+$ such that $A(0) = 0$. We suppose A differentiable, A'/A nonincreasing on $\mathbb{R}_+^*$, $\lim_{x \to +\infty} A'(x)/A(x) = 2\rho \geq 0$ and $A'(x)/A(x) = \alpha/x + B(x)$ in a neighborhood of zero, with $\alpha > 0, B$ a C^∞ odd function on $\mathbb{R}$. Let us consider the operator

$$\Delta = \frac{d^2}{dx^2} + \frac{A'(x)}{A(x)} \frac{d}{dx}. \tag{2.2.2}$$

For every C^∞ even function f on $\mathbb{R}$, the solution u on $\mathbb{R}_+ \times \mathbb{R}_+$ of the Cauchy hyperbolic problem $\Delta_x u = \Delta_y u$, with initial conditions $u(x,0) = f(x)$ and $\frac{\partial}{\partial y} u(x,0) = 0$, can be written in the form

$$u(x,y) = \int_0^{+\infty} f(t)\mu_{xy}(dt), \tag{2.2.3}$$

where $\mu_{xy} \in M_1(\mathbb{R}_+)$ is a probability measure with support the interval $[|x-y|, x+y]$ [C]. If we set $\delta_x * \delta_y = \mu_{xy}, x^- = x, e = 0, (\mathbb{R}_+, *)$ with the usual topology is called the C-T hypergroup with function A. The Haar measure is $\omega(dx) = A(x)dx$ and $\mathcal{X}(K)$ is the set of functions $\varphi_\lambda(\lambda \in \mathbb{C})$ that are solutions of the eigenvalue problem

$$\Delta\varphi_\lambda = -(\lambda^2 + \rho^2)\varphi_\lambda, \quad \varphi_\lambda(0) = 1 \text{ and } \varphi'_\lambda(0) = 1. \tag{2.2.4}$$

Moreover, the dual $\hat{\mathbb{R}_+}$ consists of all $\varphi_\lambda \in \mathcal{X}(K)$ with $\lambda \in \mathbb{R}_+ \bigcup i[0,\rho]$.

2.2.5. Polynomial hypergroups:

Let p_n, q_n and r_n be three sequences of real numbers such that $p_n > 0, r_n \geq 0, q_{n+1} > 0, q_0 = 0$ and $p_n + q_n + r_n = 1$ for all $n \in \mathbb{N}$. The polynomials defined by $P_0 \equiv 1, P_1(x) = x$ and

$$xP_n(x) = q_nP_{n-1}(x) + r_nP_n(x) + p_nP_{n+1}(x) \qquad (n \geq 1) \tag{2.2.6}$$

are orthogonal polynomials on $[-1,1]$ with respect to some measure $d\Pi(x)$. If their linearization coefficients are nonnegative (i.e. for all m, n, we have $P_m(x)P_n(x) = \sum_{r=|m-n|}^{m+n} c(m,n,r)P_r(x)$ with $c(m,n,r) \geq 0$ for all r), we can define an Hermitian hypergroup structure $(\mathbb{N}, *)$ on $\mathbb{N}$ with $e = 0$ and

$$\delta_m * \delta_n = \sum_{r=|m-n|}^{m+n} c(m,n,r)\delta_r. \tag{2.2.7}$$

It is called a polynomial hypergroup with parameters $(p_n), (q_n)$ and (r_n). The Haar measure is given by $\omega = \sum_{n=0}^{\infty} \omega(n)\delta_n$ with

$$\omega(n) = \frac{p_0p_1\cdots p_{n-1}}{q_1\cdots q_n} \quad (n \geq 1) \quad (\omega(0) = 1). \tag{2.2.8}$$

The characters are the functions on $\mathbb{N} : n \to P_n(x)$ with $x \in [-1,1]$ (see [B-H]).

2.2.9. The disk polynomial hypergroup:

Let $\alpha > 0$. The disk polynomials $(R^{\alpha}_{k,\ell}(z,\bar{z}))_{(k,\ell)\in\mathbb{N}^2}$ are orthogonal on the unit disk $D = \{z \in \mathbb{C}; |z| \leq 1\}$ with respect to the measure

$$\pi_\alpha(dxdy) = \frac{\alpha+1}{\pi}(1-x^2-y^2)^\alpha dxdy, \tag{2.2.10}$$

and are normalized by $R^\alpha_{k,\ell}(1,1) = 1$. In polar coordinates they can be expressed in terms of the Jacobi polynomials $P_i^{(\alpha,\beta)}$ (normalized by $P_i^{(\alpha,\beta)}(1) = 1$) by the formula

$$R^\alpha_{k,\ell}(\rho e^{i\theta}, \rho e^{-i\theta}) = e^{i(k-\ell)\theta}\rho^{|k-\ell|}P^{(\alpha,|m-n|)}_{k\wedge\ell}(2\rho^2-1). \tag{2.2.11}$$

By a result of Koornwinder [Ko], these polynomials satisfy a linearization formula with nonnegative coefficients:

$$R^\alpha_{k_1,\ell_1}R^\alpha_{k_2,\ell_2} = \sum_{(k,\ell)} c_\alpha(k_1,\ell_1,k_2,\ell_2;k,\ell)R^\alpha_{k,\ell}, \tag{2.2.12}$$

where the summation is over pairs (k,ℓ) such that

(2.2.13) $k_1 + k_2 + \ell = \ell_1 + \ell_2 + k$ and $|k_1 + \ell_1 - k_2 - \ell_2| \leq k + \ell \leq k_1 + \ell_1 + k_2 + \ell_2$.

We can then define a convolution on $\mathbb{N}^2$ by

$$\delta_{(k_1,\ell_1)} * \delta_{(k_2,\ell_2)} = \sum_{(k,\ell)} c_\alpha(k_1,\ell_1,k_2,\ell_2;k,\ell)\delta_{(k,\ell)}. \tag{2.2.14}$$

We then get a commutative hypergroup $(\mathbb{N}^2, *)$ with $e = (0,0)$ and involution $(k,\ell)^- = (\ell,k)$. It is called the disk polynomial hypergroup [B-G] or the dual disk hypergroup ([B-H] or [Z]).

2.3. General properties of moment functions

Let $(K,*)$ be a hypergroup. We will denote by $\mathcal{M}_1$ the set of all real valued measurable and locally bounded functions m_1 on K satisfying (1.1), i.e. $\mathcal{M}_1$ is the set of all moment functions of order 1. We denote also by $\mathcal{M}$ the set of all pairs of moment functions on K. We begin this section with some information about the size of $\mathcal{M}$. The first result is useful even if its proof is trivial.

2.3.1. Proposition. *(a) $\mathcal{M}_1$ is a real vector space (for the usual addition of functions and scalar multiplication).*

(b) If (m_1,m_2) and $(m_1,\tilde{m}_2)$ belong to $\mathcal{M}$ then $m_2 - \tilde{m}_2 \in \mathcal{M}_1$. Conversely, if $(m_1,m_2) \in \mathcal{M}$ then $(m_1, m_2 + \tilde{m}_1) \in \mathcal{M}$ for every $\tilde{m}_1 \in \mathcal{M}_1$ such that $m_1^2 \leq m_2 + \tilde{m}_1$.

We now consider the example of a topological group.

2.3.2. Proposition. *If G is a locally compact topological group, then every pair of moment functions on G is of the form (m_1, m_1^2) for some $m_1 \in \mathcal{M}_1$.*

Proof. Let $(m_1, m_2) \in \mathcal{M}$. For every $x \in G$, we have $m_1(xx^{-1}) = m_1(x) + m_1(x^{-1}) = m_1(e) = 0$ so that $m_1(x^{-1}) = -m_1(x)$. This implies, by using (1.2) with $y = x^{-1}$, the following equation:

$$m_2(x) + m_2(x^{-1}) = 2(m_1(x))^2,$$

using again $m(e) = 0$. By condition (1.3) we deduce immediately that $m_2(x) = m_2(x^{-1}) = (m_1(x))^2$. ■

2.3.3. Remark. If $G = \mathbb{R}^d$ with its additive structure and usual topology, (2.3.2) applies but more can be said. In this case $\mathcal{M}_1$ coincides with the space of continuous linear functionals on $\mathbb{R}^d$. Indeed, letting $m_1 \in \mathcal{M}_1$, we have $m_1(-x) = -m_1(x)$ and $m_1(x+y) = m_1(x)+m_1(y)$ for all x and y in $\mathbb{R}^d$. But this implies continuity of m_1 by classical arguments (recall that we have assumed measurability and local boundedness of m_1). Finally, continuity of m_1 implies $m_1(\alpha x) = \alpha m_1(x)$ for every $x \in \mathbb{R}^d$ and $\alpha \in \mathbb{R}$.

The property of having moment functions of the form (m_1, m_1^2) is in a certain sense a characteristic of groups.

2.3.4. Proposition. *Let $(K, *)$ be a hypergroup. Suppose there exists a pair of moment functions of the form (m_1, m_1^2) with $m_1 \in \mathcal{M}_1$ and $m_1(x) \neq 0$ for all $x \neq e$. Then $(K, *)$ is a topological group.*

Proof. We have $\langle \delta_x * \delta_y, m_1^2 \rangle = \langle \delta_x * \delta_y, m_1 \rangle^2$, which implies that m_1 is a constant on $\mathrm{supp}(\delta_x * \delta_y)$ for every $x, y \in K$. In particular $m_1(u) = m_1(e) = 0$ for every $u \in \mathrm{supp}(\delta_x * \delta_{x^-})$ because $e \in \mathrm{supp}(\delta_x * \delta_{x^-})$. This implies $\mathrm{supp}(\delta_x * \delta_{x^-}) = \{e\}$, i.e.:

$$(2.3.5) \qquad \delta_x * \delta_{x^-} = \delta_e \qquad (\forall x \in K).$$

Let now $x, y \in K$ and $\mu = \delta_x * \delta_y$. We have to prove that μ is a Dirac measure. By associativity of the convolution we have $\delta_{x^-} * \mu = \delta_y$, so that for every bounded measurable function f on K we have

$$(2.3.6) \qquad \langle \mu, \delta_x * f \rangle = f(y).$$

Choose f such that $f(y) = 1$ and $0 \leq f(u) < 1$ for all $u \neq y$. From (2.3.6) it follows that the function $\delta_x * f$ is equal to 1 μ almost everywhere. But $\delta_x * f(t) = 1$ means $\langle \delta_{x^-} * \delta_t, f \rangle = 1$ so $f = 1$ $\delta_{x^-} * \delta_t$ a.e. and from the special form of f we deduce that $\delta_{x^-} * \delta_t = \delta_y$. Multiplying both sides of this equation by δ_x gives $\delta_t = \delta_x * \delta_y$. ■

2.3.7. Remark. If the condition $m_1(x) \neq 0$ for every $x \neq e$ is not assumed, the result of the proposition fails because m_1 can be identically 0 on a subhypergroup of $(K, *)$. Take for example $K = G \times C$, the direct product of a topological group by a compact hypergroup, and (m_1, m_1^2) a pair of moment functions on G. Then if we set

$$\tilde{m}_1(x, y) = m_1(x)$$

for every $(x, y) \in G \times C$, it is easily verified that $(\tilde{m}_1, \tilde{m}_1^2)$ is a pair of moment functions on $(K, *)$, and $(K, *)$ is not a group. We now come to the continuity properties of moment functions.

2.3.8. Proposition. *([Z] 5.12 and 5.21): Let (m_1, m_2) be a pair of moment functions on a hypergroup $(K, *)$ with a Haar measure ω. Then m_1 and m_2 are continuous functions on K.*

Proof. Let g be a continuous and compactly supported function on K. By axioms C4, C5 and by Lebesgue's dominated convergence theorem, it is easy to see that the function

$$x \longmapsto \int_K m_1(y)(\delta_{x^-} * g)(y)\omega(dy)$$

is continuous on K. But it is also equal to

$$\int_K (\delta_x * m_1)(y) g(y)\omega(dy) = m_1(x) \int_K g(y)\omega(dy) + \int_K m_1(y) g(y)\omega(dy)$$

and the continuity of m_1 follows. A similar argument using (1.2) gives the continuity of m_2. ∎

2.3.9. Remarks

(i) A non-identically-zero first-order moment function m_1 cannot be bounded on K ([Z], 5.13). Indeed suppose $\sup_K m_1 = a < +\infty$ and $\inf_K m_1 = b > -\infty$. For every $\epsilon > 0$, choose $x_\epsilon \in K$ such that $m_1(x_\epsilon) > a - \epsilon$. Then we have

$$a \geq \langle \delta_{x_\epsilon} * \delta_y, m_1 \rangle = m_1(x_\epsilon) + m_1(y) > a - \epsilon + m_1(y)$$

for every $y \in K$. Hence $m_1(y) \leq \epsilon$ which implies $m_1 \leq 0$. A similar argument using b gives $m_1 \geq 0$. Then $m_1 \equiv 0$.

(ii) In the same manner we can show that, if $(m_1, m_2) \in \mathcal{M}$ and m_2 is bounded on K, then $m_2 \equiv 0$ and $m_1 \equiv 0$. The local boundedness of

m_1 and m_2 implies that, on a compact hypergroup, moment functions are identically zero.

(iii) Another example of a hypergroup on which moment functions are always trivial is furnished by a connected simple (nontrivial) topological group G. It is an easy exercise to prove in this case that if $m_1 \in \mathcal{M}_1$ then $m_1 \equiv 0$.

3. Determination of moment functions

In this section we suppose that $(K, *)$ is a commutative hypergroup.

3.1. Admissible paths in the structure space

Let $t \longmapsto \varphi_t$ be a map from $[0,1]$ into $\mathcal{X}(K)$ such that $\varphi_0 \equiv 1$ and the derivatives

$$m_1(x) = \frac{\partial}{\partial t}\varphi_t(x)|_{t=0} \tag{3.1.1}$$

and

$$m_2(x) = \frac{\partial^2}{\partial t^2}\varphi_t(x)|_{t=0}, \tag{3.1.2}$$

exist locally uniformly in x and are real-valued. We will then say that $t \longmapsto \varphi_t$ is an admissible path in $\mathcal{X}(K)$. By the Leibniz differentiation rule it is easy to see that m_1 and m_2 are measurable locally bounded functions on K that satisfy conditions (1.1) and (1.2).

Condition (1.3) is unfortunately not always true but there are some interesting situations in which (1.3) holds. In order to introduce them, note that typical examples of real-valued functions f on $[0,1]$ satisfying

$$(f'(0))^2 \leq f''(0) \tag{3.1.3}$$

are $f(t) = e^{\beta t}$ ($\beta \in \mathbb{R}$) and convex combinations of such functions, i.e. $f(t) = \sum_{i=1}^{n} \alpha_i e^{\beta_i t}$ with $\alpha_i \geq 0$. More generally, Laplace transforms $f(t) = \int_{\mathbb{R}} e^{\beta t}\nu(d\beta)$, where ν is for example a compactly supported positive measure, also satisfy (3.1.3) by Cauchy–Schwarz inequality. These trivial remarks apply to the cases where admissible paths in $\mathcal{X}(K)$ have a Laplace representation of the form

$$\varphi_t(x) = \int_{\mathbb{R}} e^{\alpha t}\nu_x(d\alpha), \tag{3.1.4}$$

where for each $x \in K, \nu_x$ is a compactly supported probability measure on $\mathbb{R}$. In this case $(m_1, m_2) \in \mathcal{M}$.

3.1.5. A trivial example in which a representation (3.1.4) holds is the case of $\mathbb{R}^d$. Let $m_1 \in \mathcal{M}_1$ (see 2.3.3); $\varphi_t(x) = e^{tm_1(x)}$ is an admissible path for which $\nu_x = \delta_{m_1(x)}$.

3.1.6. For C-T hypergroups (see 2.2.1) we know [T2] that there exists a continuous kernel $K(x,u)$ such that

$$(3.1.7) \qquad \varphi_\lambda(x) = \int_{-x}^{x} K(x,u)\cos(\lambda u)du \quad (\lambda \in \mathbb{C}).$$

We immediately see that

$$t \longmapsto \varphi_{i(t+\rho)}(x) = \int_{-x}^{x} K(x,u)\cosh((t+\rho)u)du$$

is an admissible path with a Laplace representation. The associated pair (m_1, m_2) of moment functions was given in ([Z] (5.17)):

$$(3.1.8) \qquad m_1(x) = 2\rho \int_0^x \int_0^y \frac{A(z)}{A(y)} dzdy$$

(3.1.9)

$$m_2(x) = 8\rho^2 \int_0^x \int_0^y \int_0^z \int_0^u \frac{A(z)A(v)}{A(y)A(u)} dvdudzdy + 2\int_0^x \int_0^y \frac{A(z)}{A(y)} dzdy.$$

3.1.4. Remark. The "path method" in $\mathcal{X}(K)$ to obtain moment functions requires a Laplace representation of the multiplicative functions. In the theory of special functions this is considered a difficult problem. In general, this method is hard to handle especially in the case of polynomial hypergroups (see the examples studied in [V]).

3.2. The case of asymptotic drift of the convolution (ADC)

Let $(K, *)$ be a commutative hypergroup with K an unbounded subset of $\mathbb{R}^d$ carrying the usual topology (the discrete topology if K is a lattice). We will suppose that the convolution is a moderate perturbation of the additive group structure; that is, it satisfies the following conditions [G2]:

(H) For some norm $|.|$ on $\mathbb{R}^d$, some constant $C > 0$ and for every $x, y \in K$ and $u \in \text{supp}(\delta_x * \delta_y)$, we have

$$|u - y| \leq C|x|.$$

Now let L be a continuous linear form on $\mathbb{R}^d$ and consider the functions on $K \times K$ defined by

$$(3.2.1) \qquad m_1(t,x) = \langle \delta_t * \delta_x, L - L(t) \rangle$$

$$(3.2.2) \qquad m_2(t,x) = \langle \delta_t * \delta_x, (L - L(t))^2 \rangle.$$

3.2.3. Definition. We say that the convolution $*$ has an asymptotic drift relative to the linear form L if the limits

$$(3.2.4) \qquad \lim_{t\to\infty} m_1(t,x) = m_1(x)$$

$$(3.2.5) \qquad \lim_{t\to\infty} m_2(t,x) = m_2(x)$$

exist for all $x \in K$ ($t \to \infty$ means $|t| \to +\infty$ and $t \in K$).

We will abbreviate this by saying that $(K,*)$ is an ADC hypergroup without reference to the linear form L when there is no risk of confusion. (This terminology is inspired by the theory of stochastic processes.) For example, condition (3.2.4) indicates that the drift in the direction L of the random walk on K with law $\mu = \delta_x$ tends to a limit at infinity; see [G1] or [G2] for details.

3.2.6. Theorem. *If $(K,*)$ is an ADC hypergroup, formulas (3.2.4) and (3.2.5) define a pair (m_1, m_2) of moment functions on K.*

The proof of this result is based on measure theoretic arguments (see [G2] for all the details). Let us only mention that property (1.3) here is trivial to obtain. Indeed by the Cauchy–Schwarz inequality we immediately get $m_1^2(t,x) \le m_2(t,x)$ for all t and $x \in K$, and $m_1^2(x) \le m_2(x)$ follows by letting $t \to \infty$.

3.2.7. Remark. The result of the theorem remains true if we suppose that t tends to infinity with some directional restrictions instead of isotropic convergence (see [G2] for details).

3.2.8. Examples

(i) In the case of $\mathbb{R}^d$, the quantities $m_1(t,x)$ and $m_2(t,x)$ are independent of t and respectively equal to $L(x)$ and $(L(x))^2$.

(ii) For the Tchebychev polynomial hypergroup (i.e. $P_n(\cos\theta) = \cos n\theta$) the convolution is given by $\delta_t * \delta_x = \frac{1}{2}(\delta_{|t-x|} + \delta_{t+x})$ for $t, x \in \mathbb{N}$. In this case $m_1(t,x) = \frac{1}{2}(|t-x| + x - t)$ ($= 0$ if $t > x$) and $m_2(t,x) = \frac{1}{2}((|t-x| - t)^2 + (t + x - t)^2)$ ($= x^2$ if $t > x$) and we have an ADC hypergroup (with $L(u) = u$). The corresponding moment functions are $m_1(x) = 0$ and $m_2(x) = x^2$.

(iii) The Bessel Kingman hypergroup is a C-T hypergroup with function $A(x) = x^{2\alpha+1}(\alpha > -1/2)$, and its convolution can be written in the form

$$(3.2.9) \qquad \langle \delta_t * \delta_x f \rangle = \int_{-1}^{1} f((t^2 + x^2 + 2txz)^{1/2}) dF_\alpha(z),$$

where $dF_\alpha(z) = \dfrac{\Gamma(\alpha+1)}{\sqrt{\pi}\Gamma(\alpha+1/2)}(1-z^2)^{\alpha-1/2}dz$ is a probability measure [Ka]. For $L(u) = u$, we have

$$(3.2.10) \qquad m_i(t,x) = \int_{-1}^{1} ((t^2 + x^2 + 2txz)^{1/2} - t)^i dF_\alpha(z) \quad (i = 1, 2).$$

It is easy to see that the limits $\lim_{t \to +\infty} m_i(t,x)$ exist and are respectively $m_1(x) = 0$ and $m_2(x) = c_\alpha x^2$ ($c_\alpha > 0$ is a constant).

3.2.11. Generalized expectation and variance

Suppose $K \subset \mathbb{R}$ and $(K, *)$ is an ADC hypergroup with moment functions (m_1, m_2) given by (3.2.4) and (3.2.5). Let us consider the associated generalized expectation $\mathbb{E}_*$ and variance V_* as defined in (1.4) and (1.5) and denote by $\mathbb{E}$ and Var the classical expectation and variance on $\mathbb{R}$. The following result gives a qualitative relation between these classical and generalized notions (see [G2] for details).

3.2.12. Proposition. *Let $\mu \in M_1(K)$ be a probability measure.*

(a) If $\mathbb{E}(\mu)$ exists then $\mathbb{E}(\delta_t * \mu)$ and $\mathbb{E}_*(\mu)$ exist ($\forall t \in K$) and

$$\mathbb{E}_*(\mu) = \lim_{t \to \infty} \mathbb{E}(\delta_t * \mu) - t.$$

(b) If $\mathrm{Var}\mu < +\infty$ then $\mathrm{Var}(\delta_t * \mu) < +\infty$ ($\forall t \in K$), $V_*(\mu)$ exists and

$$V_*(\mu) = \lim_{t \to \infty} \mathrm{Var}(\delta_t * \mu).$$

4. Explicit moment functions in the classical cases

4.1. C-T hypergroups

We determined a pair of moment functions in Section 3 by the method of admissible paths in $\mathcal{X}(K)$. The method of ADC hypergroups also works in this case. We quote the following result from [G2]:

4.1.1. Proposition. *C-T hypergroups are ADC (with respect to $L(u) = u$) and the associated moment functions m_1 and m_2 in the sense of (3.2.4) and (3.2.5) are the same as those given by (3.1.8) and (3.1.9).*

The idea of the proof is the following: Let $g_t(x) = \langle \delta_t * \delta_x, L \rangle$. It can be proved that $g_t(u) = (\delta_t * A'/A)(u)$ because Δ commutes with generalized translations. Now $\lim_{t\to\infty}(\delta_t * A'/A)(u) = 2\rho$ and we can use Lebesgue's convergence theorem as $t \to \infty$ in the integral

$$g_t(x) - t = \int_0^x \left(\int_u^x \frac{dz}{A(z)} \right) (\Delta g_t)(u) A(u) du.$$

A similar but more sophisticated argument holds to obtain m_2 (see [G2] for further details).

4.1.2. Remark. The question "Why do the two methods give the same moment functions?" is open. As noted in Proposition 2.3.1 $\mathcal{M}_1$ is a vector space; we would like to prove that $\dim \mathcal{M}_1 = 1$. Let $m_1 \in \mathcal{M}_1$. If we could prove m_1 is $\mathcal{C}^\infty, m_1(0) = m_1'(0) = 0$, then $u(x,y) = \langle \delta_x * \delta_y, m_1 \rangle$ would satisfy $\Delta_x u = \Delta_y u$ for all x, y. This would imply $\Delta m_1 = C$ (constant) and the only solution of this equation (with $m_1(0) = m_1'(0) = 0$) is $m_1(x) = C \int_0^x A(u)^{-1} \int_0^u A(\xi) d(\xi) du$. We can prove that m_1 is $\mathcal{C}^\infty$ by the following argument suggested by K. Trimèche: Let v be a $\mathcal{C}^\infty$ even function on $\mathbb{R}$ with compact support such that $\langle m, v \rangle = 1$. Then $m_1 * v$ is of class $\mathcal{C}^\infty$ and we have

$$m_1 * v(x) = \int_0^\infty (\delta_x * m_1)(y) v(y) A(y) dy = m_1(x) + \int_0^\infty v(y) m_1(y) A(y) dy.$$

This clearly proves the result.

4.2. Polynomial hypergroups

Let $(\mathbb{N}, *)$ be a polynomial hypergroup as in Section 2.2.5. Let us denote $P_m(x) = \sum_{k=0}^{m} \alpha_k^{(m)} x^k$. From the definition of the convolution we deduce the crucial representation

$$\delta_m = P_m(\delta_1) \qquad (\forall m \in \mathbb{N}), \tag{4.2.1}$$

where in the expression of P_m, x^k is replaced by $\delta_1^k = \delta_1 * \cdots * \delta_1$ (k times). Let us begin with an easy fact concerning the derivative of the polynomials P_n at the point 1.

4.2.2. Proposition. *$(P_n'(1))_{n\geq 0}$ is a strictly increasing sequence of positive numbers.*

Proof. From (2.2.6) we immediately deduce that the sequence $u_n = P_n'(1)$ satisfies

$$(4.2.3) \qquad 1 + u_n = p_n u_{n+1} + r_n u_n + q_n u_{n-1} \qquad (n \geq 1),$$

with $u_0 = 0$ and $u_1 = 1$. We can rewrite (4.2.3) in the form

$$(4.2.4) \qquad 1 + q_n(u_n - u_{n-1}) = p_n(u_{n+1} - u_n),$$

and the result follows clearly by induction. ■

4.2.5. Proposition. *If a polynomial hypergroup has a pair (m_1, m_2) of moment functions, then there exist $a \in \mathbb{R}$ and $b \geq 0$ such that*

$$(4.2.6) \qquad m_1(n) = aP_n'(1)$$

$$(4.2.7) \qquad m_2(n) = a^2 P_n''(1) + bP_n'(1).$$

Conversely, if there exist constants $a \in \mathbb{R}$ and $b \geq 0$ such that

$$(4.2.8) \qquad a^2 P_n''(1) + bP_n'(1) \geq a^2 (P_n'(1))^2$$

for all $n \in \mathbb{N}$, then the hypergroup has moment functions given by (4.2.6) and (4.2.7).

Proof. Let (m_1, m_2) be a pair of moment functions. By (1.1) and iteration, for every $k \in \mathbb{N}^*$ and $y \in \mathbb{N}$, we have

$$(\delta_1^k * m_1)(y) = \langle \delta_1^k * \delta_y, m_1 \rangle = km_1(1) + m_1(y).$$

We deduce that, for every polynomial $P(x) = \sum_{k=1}^{n} \alpha_k x^k$,

$$\begin{aligned} (P(\delta_1) * m_1)(y) &= \sum_{k=1}^{n} \alpha_k (\delta_1^k * m_1)(y) \\ &= P'(1) m_1(1) + P(1) m_1(y). \end{aligned}$$

With $P = P_n$ and using (4.2.1) we immediately obtain (4.2.6) with $a = m_1(1)$. To get (4.2.7) consider the associated variance V_* given by (1.5). For every $k \in \mathbb{N}$, we have

$$(4.2.9) \qquad V_*(\delta_1^k) = (\delta_1^k * m_2)(0) - k^2 m_1^2(1) = kV_*(\delta_1).$$

Multiplying this relation by α_k and summing over $k = 1, \cdots, n$, we obtain

$$(P(\delta_1) * m_2)(0) - (P''(1) + P'(1))m_1^2(1) = P'(1)V_*(\delta_1),$$

where $P(x) = \sum_{k=1}^{n} \alpha_k x^k$. If $P = P_n$, using (1.2) and (4.2.1), and taking into account that $V_*(\delta_1) = m_2(1) - m_1^2(1)$ and $m_1(0) = 0$, we immediately get (4.2.7) (with $a = m_1(1)$ and $b = m_2(1)$).

For the converse, let m_1 and m_2 be defined by (4.2.6) and (4.2.7). Assumption (4.2.8) is condition (1.3). Let us verify (1.1). For every $y \in \mathbb{N}$ we have

$$\begin{aligned}(\delta_1 * m_1)(y) &= p_y m_1(y+1) + r_y m_1(y) + q_y m_1(y-1) \\ &= a(P_y'(1) + 1) = m_1(y) + a.\end{aligned}$$

By induction we obtain $(\delta_1^k * m_1)(y) = m_1(y) + ka$, and this implies that $(\delta_n * m_1)(y) = (P_n(\delta_1) * m_1)(y) = m_1(y) + \left(\sum_{k=0}^{n} \alpha_k^n k\right) a = m_1(y) + m_1(k)$. The same arguments also work to verify condition 1.2. ■

4.2.9. Remark. Condition (4.2.8) is always satisfied with $a = 0$ and $b > 0$ but in this case $m_1 \equiv 0$. The pair (m_1, m_2) with $m_1 \equiv 0$ will be called a trivial pair of moment functions. The following result is a direct consequence of 4.2.5.

4.2.10. Corollary. *A polynomial hypergroup has a non-trivial pair of moment functions if and only if*

$$\sup_{n>0} \frac{P_n'(1)^2 - P_n''(1)}{P_n'(1)} < +\infty.$$

In this case with the notations of (2.2.1) we have $\dim \mathcal{M}_1 = 1$ *and the uniqueness of* m_2 *modulo* $\mathcal{M}_1$ *is also guaranteed.*

4.2.11 Example

The Tchebychev polynomial hypergroup (i.e. with $P_n(\cos\theta) = \cos n\theta$) has only trivial moment functions (i.e. $m_1 \equiv 0$ and $m_2(n) = bP_n'(1)$). Indeed, we have $P_n'(1) = n^2$ and $P_n''(1) = 1/3(n^4 - n^2)$, and the sequence $P_n'(1)^{-1}(P_n'(1)^2 - P_n''(1))$ tends to infinity.

4.2.12. Remark. Explicit expressions in terms of the parameters p_n, q_n, r_n for the sequences $P_n'(1)$ and $P_n''(1)$ are not difficult to obtain.

If we differentiate (2.2.6) we see that the sequences $u_n = P_n'(1)$ and $v_n = P_n''(1)$ satisfy the recurrence relations (4.2.3) and

$$(4.2.13) \qquad 2u_n + v_n = p_n v_{n+1} + r_n v_n + q_n v_{n-1} \qquad (v_0 = v_1 = 0).$$

If we let $\Delta_k = u_k - u_{k-1}$, we get from (4.2.3) the linear difference equation

$$(4.2.14) \qquad \Delta_k = a_{k-1}\Delta_{k-1} + b_{k-1},$$

where $a_k = q_k/p_k$ and $b_k = 1/p_k$. By finite induction we obtain

$$(4.2.15) \qquad \Delta_k = \sum_{j=0}^{k-1}\left(b_j \prod_{i=j+1}^{k-1} a_i\right),$$

and this immediately determines u_n by summing Δ_k over $k = 1, \cdots, n$. In the same manner $\Delta_k' = v_k - v_{k-1}$ satisfies an equation like (4.2.14) with $a_k = q_k/p_k$ and $b_k = 2u_k/p_k$, and the result is analogous to (4.2.15); we omit the details.

We will now give an example of a class of polynomial hypergroups with asymptotic drift of the convolution. We will say that a polynomial hypergroup has converging parameters if the limits $\lim_{n\to\infty} p_n = p \in]0,1[$ and $\lim_{n\to\infty} q_n = q \in]0,1[$ exist.

4.2.16. Proposition. *A polynomial hypergroup with converging parameters is an ADC hypergroup (with $L(u) = u$), and the moment functions according to (3.2.4) and (3.2.5) are given by*

$$(4.2.17) \qquad m_1(n) = (p-q)P_n'(1)$$

$$(4.2.18) \qquad m_2(n) = (p+q)P_n'(1) + (p-q)^2 P_n''(1).$$

Proof. (see [G2] for all details): We will give only the main ideas. We have $\langle \delta_x * \delta_1, L - x\rangle = p_x - q_x \to p - q$ as $x \to +\infty$. By induction, for every $k \in \mathbb{N}$, we obtain

$$\lim_{x\to+\infty} \langle \delta_x * \delta_1^k, L - x\rangle = k(p-q).$$

Then using (4.2.1) we deduce

$$\lim_{x\to\infty} \langle \delta_x * \delta_n, L - x\rangle = (p-q)P_n'(1) = m_1(n).$$

In the same way we can prove that

$$\lim_{x\to\infty}\langle\delta_x*\delta_1^k,(L-x)^2\rangle = k(p+q)+k(k-1)(p-q)^2$$

and using (4.2.1) again we get the formula for $m_2(n)$. ■

Remark. A polynomial hypergroup with converging parameters such that $p > q$ has nontrivial moment functions. Using Corollary 4.2.10, this shows indirectly that the condition $\sup_{n>0}(P_n'(1))^{-1}(P_n'(1)^2 - P_n''(1)) < +\infty$ is satisfied. We don't know if this is easy to verify by analytical methods.

4.2.20. Remark. From formula (4.2.15) and the expression for the Haar measure given in (2.2.8) we can rewrite $P_n'(1)$ in the form

$$P_n'(1) = \sum_{k=1}^{n}\sum_{j=0}^{k-1}\frac{\omega(j)}{\omega(k)q_k} = \sum_{k=1}^{n}\frac{\omega([0,k-1])}{\omega(k)q_k}. \tag{4.2.21}$$

This establishes a link between $m_1(n)$ and the Haar measure analogous to the case of C-T hypergroups where we have found $m_1(x) = 2\rho\int_0^x \frac{1}{A(y)}\omega([0,y])dy$ (see (3.1.8)). This remark is a very incomplete answer to a question of Prof. Ajit Iqbal Singh concerning the relations between moment functions and Haar measure. This remains an interesting and open problem.

4.3. Moment functions on the disk polynomial hypergroup

From the structure of the support of $\delta_{(m,n)}*\delta_{(k,\ell)}$ (see (2.2.13) and (2.2.14)) we can see easily that the linear form $L(u,v) = u-v$ is constant on $\mathrm{supp}(\delta_{(m,n)}*\delta_{(k,\ell)})$. This shows that the functions $m_1(t,x)$ and $m_2(t,x)$ do not depend on t and so the convolution has asymptotic drift relative to L with moment functions

$$m_1(k,\ell) = k-\ell \text{ and } m_2(k,\ell) = (k-\ell)^2. \tag{4.3.1}$$

We quote without proof the following result from [G2], which gives another pair of moment functions.

4.3.2. Proposition. *Convolution on* $(\mathbb{N}^2,*)$ *has asymptotic drift relative to the linear form* $L(u,v) = u+v$ *when* $t\to\infty$ *in the direction of the half line* $u = v$ *(see Remark 3.2.7). The corresponding moment functions are given by*

$$m_1(k,\ell) = 0 \text{ and } m_2(k,\ell) = \tfrac{2k\ell}{\alpha+1}+k+\ell. \tag{4.3.3}$$

4.3.4. Remarks

(i) Moment functions (4.3.3) were used in [B-G] to obtain a law of large numbers on $(\mathbb{N}^2, *)$.

(ii) The problem of determining all possible pairs of moment functions on the polynomial disk hypergroupis open, but in our opinion a generalization of the methods used in Section 4.2 to multidimensional polynomial hypergroups is possible.

References

[B-G] BOUHAIK, M., GALLARDO, L. [1992] Un théorème limite central dans un hypergroupe bidimensionnel. Ann. Inst. Henri Poincaré Prob. Stat. **28** $n°$ 1, p. 47–61.

[B-H] BLOOM, W., HEYER, H. [1995] Harmonic analysis of probability measures on hypergroups. Walter de Gruyter Ed., Berlin-New York.

[C] CHEBLI, H. [1974] Opérateurs de translation généralisée et semi groupes de convolution. Lecture Notes in Math. **404**, p. 35–59.

[F] FARAUT, J. [1975] Dispersion d'une mesure de probabilité sur $SL(2, \mathbb{R})$ biinvariante par $SO(1, \mathbb{R})$ et théorème de la limite centrale. Université Louis Pasteur - Strasbourg.

[G1] GALLARDO, L. [1996] Chaînes de Markov à dérive stable et lois des grands nombres sur les hypergroupes. Ann. Inst. H. Poincaré, Prob. Stat., à paraître.

[G2] GALLARDO, L. [1996] Asymptotic drift of the convolution and moment functions on hypergroups. Math. Zeitschrift, to appear.

[Ka] KARPELEVICH, F.I., TUTUBALIN, V.N., SHUR, M.G. [1959] Limit theorems for the composition of distributions in the Lobachevsky plane and space. Theory of Prob. and Appl. **4**, p. 399–402.

[Ko] KOORWINDER, T.H. [1978] Positivity proofs for linearization and connection coefficients of orthogonal polynomials satisfying an addition formula. J. London Math. Soc. **28**, p. 101–114.

[Ki] KINGMAN J.F.C. [1963] Random walks with spherical symmetry. Acta Math. **109**, p. 11–53.

[S] SPECTOR, R. [1978] Mesures invariantes sur les hypergroupes. Trans. Amer. Soc. **239**, p. 147–165.

[T1] TRIMECHE, K. [1978] Probabilités indéfiniment divisibles et théorème de la limite centrale pour une convolution généralisée sur la demi droite. Comptes Rendus Acad. Sc. Paris, Série A, t. **286**, p. 63–66.

[T2] TRIMECHE, K. [1981] Transformation intégrale de Weyl et théorème de Paley-Wiener associés à un opérateur différentiel singulier sur $[0,\infty[$. J. Math. Pures-Appl. **60**, p. 51–98.

[V] VOIT [1990] Laws of large numbers for polynomial hypergroups and some applications. J. Theor. Prob. **3**, p. 245–266.

[Z] ZEUNER, H. [1992] Moment functions and laws of large numbers on hypergroups. Math. Zeitschrift **211**, p. 369–407.

Université de Brest, Département de Mathématiques, 6 Avenue Le Gorgeu BP 809 29285 BREST - FRANCE.

From September 1st 1996: Université de Tours, Département de Mathématiques, Faculté des Sciences et Techniques, Parc de Grandmont 37200 TOURS - FRANCE

About Some Random Fourier Series and Multipliers Theorems on Compact Commutative Hypergroups

Marc-Olivier Gebuhrer

Abstract

The character theory of compact commutative hypergroups is yet far from being well understood; as this paper shows, the behaviour of the Plancherel measure is related to some deeper harmonic analysis involving notably Sidon sets. Pioneering work related to this area has been performed a long time ago by R. Vrem, K. Ross, J. Fournier (see references below). This elementary study provides only a sample of easy results, but paves the way to apparently much deeper questions.

This paper is comprised of two parts: the first one has been already mentioned; the second provides some easy multiplier theorems. In the context of hypergroups, and despite its elementary character, much of the material presented here is new.

1. General notations; preliminary results

Let X be a compact topological Hausdorff space; we denote by $M(X)$ the Banach space of complex valued Radon measures on X, endowed with its natural dual norm $\|\mu\|_1 = \sup\{|\langle\mu, f\rangle|; \|f\|_\infty \leq 1, f \in C(X)\}$ where $C(X)$ stands for the Banach space of complex valued continuous functions on X.

The following definition recalls our notion of compact commutative hypergroup as given in $[G]_1$.

Definition 1.1. Let us assume that $(M(X), \|\cdot\|_1)$ is endowed with a commutative Banach algebra structure. We denote the product on $M(X)$ by $*$ and call it thereafter convolution on $M(X)$.

Assume further that an involutive homeomorphism $x \mapsto \check{x}$ is given on X. We shall say that $(M(X), *, \vee)$ is a commutative hypergroup on the compact space X provided the following requirements are fulfilled:

(NA_1) The mapping $(x, y) \mapsto \delta_x * \delta_y$ from $X \times X$ into $M(X)$ is w^*-continuous.

(NA_2) The convex set $M^1(X)$ of probability measures on X is a semigroup for the convolution.

(U_1) There exists a unique point $e \in X$ such that δ_e is the unit of $M(X)$ and moreover $e \in \operatorname{Supp}(\delta_x * \delta_{\check{y}}) \Leftrightarrow x = y$.

(U_2) For each $x, y \in X$ such that $x \neq y$, there exist neighborhoods $W(x), W(y)$ of x, y respectively such that $e \notin \operatorname{clos}(W(x)*W(\check{y}))$ (where clos denotes the closure in X of the bracketed set, and where, as usual, for two subsets A, B of X, we denote $A * B := \bigcup_{\substack{a\in A\\ b\in B}} \operatorname{Supp}(\delta_a * \delta_b)$.

1.2. We shall denote by σ the unique translation invariant probability measure on X. Let us recall that for $f \in C(X)$, we let

$$\tau_x f(y) = \langle \delta_{\check{x}} * \delta_y, f\rangle \text{ and that } \langle \tau, \tau_x f\rangle = \langle \tau, f\rangle \text{ for } x \in X,\ f \in C(X).$$

It is well known that $\operatorname{Supp}(\tau) = X$.

Let us also recall that $L^1(X, \sigma)$ is a closed ideal in $M(X)$ and that the following property holds:

$$(f.\sigma) * (g.\sigma) = h.\sigma$$

where

$$h(x) := \int_X f(y)\tau_y g(x) d\sigma(y) = \int_X g(y)\tau_y f(x) d\sigma(y) \text{ for } f, g \in L^1(X, \sigma)$$

the former equalities being understood almost everywhere w.r.t. σ. We denote by $\hat{X}$ the spectrum of the commutative Banach algebra $L^1(X, \sigma)$, which is known to be symmetric, X being compact [S].

It is also classical that for the natural topology on the Gel'fand spectrum of the Banach algebra $L^1(X, \sigma)$, $\hat{X}$ is a discrete topological space; [B-H], [S]. The complex vector space generated by elements of $\hat{X}$, viewed as continuous functions on X, will be called the space of trigonometric polynomials, and denoted by $P(X)$. We define the Fourier transform of $f \in L^1(X, \sigma)$ by the formula:

$$\hat{f}(\chi) := \int_X f(x)\overline{\chi}(x) d\sigma(x).$$

The Fourier transform extends to measures on X in the obvious way.

We shall use repeatedly the following fundamental results:

Theorem 1.3. (Plancherel). *The Fourier transform $f \mapsto \hat{f}$ restricted to $L^2(X, \sigma)$ yields an isometric isomorphism from $L^2(X, \sigma)$ onto $\ell^2(\hat{X}, \varpi)$ where ϖ is the Plancherel measure on $\hat{X}$ defined by $\varpi\{\chi\} = \|\chi\|_2^{-2}$ for $\chi \in \hat{X}$.*

Theorem 1.4. (Fèjer approximation). *There exists a bounded approximate unit* $\{\varphi_\alpha\}_{\alpha \in A}$ *in* $L^1(X,\sigma)$ *such that*

(i) for each $\alpha \in A$, $\varphi_\alpha \in P(X)$,

(ii) for any $\varepsilon > 0$, *there exists a cofinal subset* $B(\varepsilon) \subset A$ *such that*

$$\|\varphi_\alpha\|_1 \leq 1 + \varepsilon \text{ for any } \alpha \in B(\varepsilon).$$

The proof of these two results may be found in [B-H] for 1.3 and $[G]_2$ for 1.4. The reader might also be interested in the forthcoming paper $[G - Sc]_2$ where the whole theory will be detailed within our axiomatics.

Theorem 1.4 has also been proved previously in a completely different way by$[V]_1$. The methods cannot be compared; the approximate identities thus constructed are different. $[G]_2$ provides an effective and universal way to compute such an approximate identity on a compact hypergroup, and the methods allow generalizations to the locally compact case. $[V]_1$ is limited to the compact case and cannot be further extended.

1.5. Standing assumption for the rest of the paper.

From now on we assume the space X to be metrisable.

1.6. Abuse of language.

We shall speak of the hypergroup X instead of $(M(X), *, \vee)$.

2. Changing signs of Fourier coefficients

We begin by mentioning the following

Theorem 2.1. (Hausdorff-Young). *Let* $p \in [1,2]$, p' *such that* $\frac{1}{p} + \frac{1}{p'} = 1$.

(i) If $f \in L^p(X,\sigma)$ *then* $\hat{f} \in \ell^{p'}(\hat{X},\varpi)$ *and* $\|\hat{f}\|_{p'} \leq \|f\|_p$.

(ii) If $\phi \in \ell^p(\hat{X},\varpi)$, *there exists* $\Phi \in L^{p'}(X,\sigma)$ *such that*

$$\hat{\Phi} = \phi \text{ and } \|\Phi\|_{p'} \leq \|\phi\|_p.$$

Moreover the series

$$\sum_n \phi(\chi_n)\chi_n\varpi\{\chi_n\} \text{ converges in } L^{p'}(X,\sigma) \text{ to } \Phi.$$

Proof. It follows the classical lines and has been proved previously in general by $[V]_2$. Just notice that by defining the linear mapping T by

$$Tf = \hat{f} \text{ for } f \in L^1(X,\sigma)$$

and working with $L^1(X,\sigma)$ and $\ell^\infty(\hat{X},\varpi)$, it is plain that T is of type $(1,\infty)$. By 1.3, T is of type (2,2). The result follows by applying the Riesz-Thorin interpolation theorem. ∎

Remark 2.1.1.

By 2.1, for $p \in [1,2]$, the linear mapping $f \mapsto \hat{f}$ is continuous from L^p into $\ell^{p'}$; in other words $\|\hat{f}\|_{p'} \leq A_p\|f\|_p$ for some positive $A_p \leq 1$. We do not know the best possible estimate for A_p. We do not know how A_p depends on the hypergroup. We do not know which functions in L^p are extremal with respect to that property.

Theorem 2.2. *Let ϕ a complex valued function on $\hat{X}$ and let $p \in [1,\infty]$.*

Let us call $C_p(\phi)$ the following property:

$$C_p(\phi) : \sum_n |\phi(\chi_n)\hat{f}(\chi_n)|\varpi\{\chi_n\} < +\infty \textit{ for every } f \in L^p(X,\sigma).$$

Then
(i) $C_1(\phi)$ holds if and only if $\phi \in \ell^1(\hat{X},\varpi)$.
(ii) If $p \in [1,2]$ and if $C_p(\phi)$ holds, then $\phi \in \ell^2(\hat{X},\varpi)$.
(iii) If $p \in [2,\infty]$ and if $\phi \in \ell^2(\hat{X},\varpi)$ then $C_p(\phi)$ holds.
(iv) If $p \in]1,2[$ and if $\phi \in \ell^p(\hat{X},\varpi)$ then $C_p(\phi)$ holds.

Proof. In what follows property $C_p(\phi)$ will be abbreviated to C_p.

(i) Let $\hat{g}$ be a complex valued function on $\hat{X}$ such that $\lim_{\chi_n\to\infty}|\hat{g}(\chi_n)| = 0$.

We denote by $c_0(\hat{X})$ the space of such functions and endow it with its natural Banach space norm $\|\hat{g}\|_\infty = \sup_n |\hat{g}(\chi_n)|$.

The linear functional $T_{\phi\hat{g}}$ defined on $L^1(X,\sigma)$ by the formula

$$\langle T_{\phi\hat{g}}, f\rangle := \sum_n \phi(\chi_n)\hat{g}(\chi_n)\hat{f}(\chi_n)\varpi\{\chi_n\}$$

is bounded on $L^1(X,\sigma)$ for any $\hat{g} \in C_0(\hat{X})$; in fact, let $\langle T^N_{\phi\hat{g}}, f\rangle := \sum_{n\leq N}\phi(\chi_n)\hat{g}(\chi_n)\hat{f}(\chi_n)\varpi\{\chi_n\}$. We see at once that the linear functionals $T^N_{\phi\hat{g}}$ are bounded on $L^1(X,\sigma)$ and assumption (C_1) tells us that $\lim_{N\to\infty}\langle T^N_{\phi\hat{g}}, f\rangle$ exists for any $f \in L^1(X,\sigma)$. Our claim follows from the Banach Steinhaus theorem. But in fact, (C_1) entails that $\sup_{\|\hat{g}\|_\infty\le 1} |\langle T_{\phi\hat{g}}, f\rangle| < +\infty$ for any $f \in L^1(X,\sigma)$, as one immediately checks; by applying the uniform boundedness principle, we see that $\sup_{\substack{\|\hat{g}\|_\infty\le 1\\ \hat{g}\in C_0(\hat{X})}} \|T_{\phi\hat{g}}\| < +\infty$ where, as usual, $\|T\|$ stands for the norm of the bounded linear functional T on $L^1(X,\sigma)$.

Let N be a fixed integer; we choose $\hat{g} \in C_0(\hat{X})$, $\|\hat{g}\|_\infty \le 1$ such that

$$\phi(\chi_n)\hat{g}(\chi_n) = |\phi(\chi_n)|$$

for $n \le N$ (as usual if $\phi(\chi_n) = 0$, we let $\hat{g}(\chi_n) = 0$); now, in statement 1.4, it is always possible to replace the given bounded approximate identity by a bounded approximate identity (K_k) with the same properties, satisfying moreover the condition $\hat{K}_k \ge 0$ on $\hat{X}$; for this, its sufficient to let $K_k = \varphi_k * \tilde{\varphi}_k$, where for any f we define $\tilde{f}(x) = \overline{f(\check{x})}(x \in X)$.

This choice being done, one has

$$\sum_{n\le N} |\phi(\chi_n)|\hat{K}_k(\chi_n)\varpi\{\chi_n\} \le |\sum_{n} \phi(\chi_n)\hat{g}(\chi_n)\hat{K}_k(\chi_n)\varpi\{\chi_n\}|$$

where the right written term is $|\langle T_{\phi\hat{g}}, K_k\rangle|$.

Therefore

$$\sum_{n\le N} |\phi(\chi_n)|\hat{K}_k(\chi_n)\varpi\{\chi_n\} \le \sup_{\substack{\|\hat{g}\|_\infty\le 1\\ \hat{g}\in C_0(\hat{X})}} \|T_{\phi\hat{g}}\| \; \|K_k\|_1.$$

And finally as $\|K_k\|_1 \le 1 + 1/k$, one has

$$\sum_{n\le N} |\phi(\chi_n)|\hat{K}_k(\chi_n)\varpi\{\chi_n\} \le \sup_{\substack{\|\hat{g}\|_\infty\le 1\\ \hat{g}\in C_0(\hat{X})}} \|T_{\phi_{\hat{g}}}\|(1 + 1/k).$$

By letting $k \to \infty$, we get

$$\sum_{n\le N} |\phi(\chi_n)|\varpi\{\chi_n\} \le \sup\{\|T_{\phi\hat{g}}\| \,;\, \|\hat{g}\|_\infty \le 1, \hat{g} \in C_0(\hat{X})\}.$$

The right hand term is now independent of N and this shows that $\phi \in \ell^1(\hat{X},\varpi)$.

(ii) $p \in [1,2]$ and (C_p) holds. Cases $p = 1,2$ are trivial; case $p = 1$ results from (i) and the fact that $\ell^1(\hat{X},\varpi) \subset \ell^2(X^1,\varpi)$. Case $p = 2$

results from the fact that $L^2(X,\sigma)$ is a Hilbert space and Theorem 1.3 (Plancherel). Let $p \in]1,2[$. By using the same device as in (i), one sees immediately that the linear functional $\langle T_\phi, f\rangle = \sum_n \phi(\chi_n)\hat{f}(\chi_n)\varpi\{\chi_n\}$ is bounded on $L^p(X,\sigma)$.

Therefore,

$$|\langle T_\phi, f\rangle| \leq C_p \|f\|_p \text{ for some positive } C_p, f \in L^p(X,\sigma).$$

But $L^2(X,\sigma) \subset L^p(X,\sigma)$ for $p \in [1,2]$ and $\|f\|_p \leq \|f\|_2$ for $f \in L^2(X,\sigma)$. Therefore, T_ϕ defines a bounded functional on $L^2(X,\sigma)$. There exists a unique $\Phi \in L^2(X,\sigma)$ such that $\int_X f(x)\Phi(x)d\sigma(x) = \langle T_\phi, f\rangle$ for $f \in L^2(X,\sigma)$. It is plain that $\hat{\Phi} = \phi$ and by 1.3 again $\phi \in \ell^2(\hat{X},\varpi)$.

(iii) If $p \in [2,\infty]$ and if $\phi \in \ell^2(\hat{X},\varpi)$, then (C_p) holds: in fact, in that case $L^p(X,\sigma) \subset L^2(X,\sigma)$; so by 1.3 again the series $\sum_n \phi(\chi_n)\hat{f}(\chi_n)\varpi\{\chi_n\}$ is absolutely convergent for any $f \in L^p(X,\sigma)$ and this is the content of (C_p).

(iv) If $p \in]1,2[$ and $\phi \in \ell^p(\hat{X},\varpi)$ then (C_p) is true. In fact by Theorem 2.1, there exists a $\Phi \in L^{p'}(X,\sigma)$ $\left(\frac{1}{p} + \frac{1}{p'} = 1\right)$ such that $\hat{\Phi} = \phi$; so consider the linear functional $f \mapsto \int_X \Phi(x)f(x)d\sigma(x)$; we know that the series $\sum_n \phi(\chi_n)\overline{\chi}_n\varpi\{\chi_n\}$ converges in $L^{p'}(X,\sigma)$ again by 2.1, so that

$$\int_X \hat{\Phi}(x)f(x)d\sigma(x) = \sum_n \phi(\chi_n)\hat{f}(\chi_n)\varpi\{\chi_n\},$$

the right term being convergent for every $f \in L^p(X,\sigma)$.

Now, again by 2.1, $\hat{f} \in \ell^{p'}(\hat{X},\varpi)$, so that the right hand series is in fact absolutely convergent for any $f \in L^p(X,\sigma)$. This is the content of (C_p). ■

2.3. Given the preceding result, the following question is in order:

Let $p \in]2,\infty]$: does property $C_p(\phi)$ imply $\phi \in \ell^2(\hat{X},\varpi)$? The answer is known to be positive for any commutative compact group G, but the proof is much deeper [H]. The answer for compact commutative hypergroups is yet unknown, but important steps towards the answer will be found in the forthcoming paper [G]. Instead of turning towards that problem, we move on to explore another, intimately connected area of problems.

Definition 2.4. The Rademacher functions $(r_m)_{m\in\mathbf{N}}$ are the functions defined on the interval $[0,1]$ by the formulae $r_m(t) =$

sgn sin $2^{n+1}\pi t (m \in \mathbf{N})$.

Classical properties of the sequence (r_m) are to be found *e.g* in [Z]. They will be used here without proof.

2.5. Let $\phi \in \ell^2(\hat{X}, \varpi)$; for $t \in [0,1]$ we let

$$f_t(x) := \sum_n r_n(t)\phi(\chi_n)\chi_n(x)\varpi\{\chi_n\}.$$

The meaning of the right-hand side series is not completely trivial.

First of all, for a.e. $t \in [0,1]$, $f_t(x) \in L^2(X,\sigma)$ and the series converges in $L^2(X,\sigma)$ to f_t; moreover, at least formally, for the moment,

$$\int_0^1 f_t(x) r_m(t)dt = \phi(\chi_m)\chi_m(x)\varpi\{\chi_m\};$$

it is not immediately clear that for fixed $x \in X$, $t \mapsto f_t(x) \in L^2([0,1],dt)$ and this is definitely not true in general for $x = e$.

However, let $\Gamma(x) := \sum_m |\phi(\chi_m)|^2 |\chi_m(x)|^2 \varpi^2\{\chi_m\}$ for $x \in X$, the sum being finite or not; we have

$$\int_X \Gamma(x)d\sigma(x) = \sum_m |\phi(\chi_m)|^2 \int_X |\chi_m(x)|^2 d\sigma(x)\varpi^2\{\chi_m\} = \|\phi\|_2^2,$$

because $\varpi\{\chi_m\} = \|\chi_m\|_2^{-2}$ so that $\Gamma(x) < +\infty$ for almost every $x \in X$ with respect to σ. Moreover

$$\int_0^1 |f_t(x)|^2 dt = \sum_m |\phi(\chi_m)|^2 |\chi_m(x)|^2 \varpi^2\{\chi_m\} \text{ for almost every } x \in X,$$

with respect to σ. Much more is true in fact.

Theorem 2.6. *Let $\phi \in \ell^2(\hat{X}, \varpi)$. For almost every $t \in [0,1]$, the series*

$$f_t(x) = \sum_n r_n(t)\phi(\chi_n)\chi_n(x)\varpi\{\chi_n\}$$

converges almost everywhere on X with respect to σ.

Proof. Let $\Gamma = \{(x,t) \in X \times [0,1]; f_t(x) \text{ is convergent}\}$; Γ is $\sigma \otimes m$ measurable in $X \times [0,1]$ where m denotes Lebesgue measure on $[0,1]$. By 2.5 there exists a subset $E_0 \subset X$, with $\sigma(E_0) = 0$, such that for every

$x_0 \in X \backslash E_0$, $m\{t \in [0,1]; \Gamma \cap \{(x_0, t)\} \neq \emptyset\} = 1$. In fact, we showed in 2.5 that for such an x_0, the series $\sum_n |\phi(\chi_n)|^2 |\chi_n(x_0)|^2 \varpi^2\{\chi_n\} < +\infty$ and the assertion is just a classical result for Rademacher functions ([Z], 5.5 (i)). So, the measure of Γ relative to $\sigma \otimes m$ is 1; therefore, by Fubini's theorem, $\sigma\{\Gamma \cap \{(x, t_0)\} \neq \emptyset\} = 1$ for almost every t_0 and this is our claim. ■

2.6.1. Remark.

Theorem 2.6 is just an extension of the classical case as found in [Z, 5.6 (i)]. However the result is relatively unexpected given the various possible growths of the Plancherel measure ϖ.

2.7. At this point, we make the following elementary observation: starting with $\phi \in \ell^2(\hat{X}, \varpi)$, the series $\sum_n |\phi(\chi_n)|^2 |\chi_n(x)|^2 \varpi\{\chi_n\}$ converges for every $x \in X$ and even,

$$\sum_n |\phi(\chi_n)|^2 |\chi_n(x)|^2 \varpi\{\chi_n\} \leq \|\phi\|_2^2$$
$$= \sum_n |\phi(\chi_n)|^2 \varpi\{\chi_n\} \text{ for every } x \in X ,$$

while, at the same time,

$$\sum_n |\phi(\chi_n)|^2 |\chi_n(x)|^2 \varpi^2\{\chi_n\} < +\infty \text{ for almost every } x \in X.$$

The exceptional set in the last written series is of course dependent on ϕ. Two natural cases present themselves (we are going to give some results for the easiest one) namely

Case (A) $\sup_n \varpi\{\chi_n\} < +\infty$

Case (B) $\sup_n \varpi\{\chi_n\} = +\infty$.

Proposition 2.8. *If* $\sup_n \varpi\{\chi_n\} = +\infty$, *then, for every neighborhood* V *of* e *in* X, *one has* $\inf_{n, x \in V} |\chi_n(x)| = 0$.

Proof. Assume the contrary and fix some neighborhood V of e in X for which $\inf_{n, x \in V} |\chi_n(x)| = \alpha > 0$. Aside from a negligible set in V, the series $\sum_n |\phi(\chi_n)|^2 |\chi_n(x)|^2 \varpi^2\{\chi_n\}$ converges. As $|\chi_n(x)| \geq \alpha$ for any n, this shows that $\sum_n |\phi(\chi_n)|^2 \varpi^2\{\chi_n\} < +\infty$ whenever $\phi \in \ell^2(\hat{X}, \varpi)$ and this clearly implies $\sup_n \varpi\{\chi_n\} < \infty$. ■

Remark 2.8.1.

Proposition 2.8 is just a slight improvement of the following immediate consequence from assumption B): as $\varpi\{\chi_n\} = \|\chi_n\|_2^{-2}$,

$\sup_n \varpi\{\chi_n\} = +\infty$ implies that $\underline{\lim}_{n\to\infty} \|\chi_n\|_2 = 0$, so that $\underline{\lim}_{n\to\infty} \chi_n(x) = 0$ almost everywhere with respect to σ.

2.9. It is time now to discuss some problems:

(i) Case (A) is the closest to the group case for which $\varpi\{\chi_n\} = 1$ for any n. We can make the question precise in the following way.

Let G be a semisimple compact connected Lie group. Describe the Gel'fand pairs (G, K) such that the hypergroup $M(K\backslash G/K)$ falls into case (A).

(ii) If the hypergroup $(M(X), *, \vee)$ is such that the involution is trivial then its characters are real functions. Because of the orthogonality relations, any non trivial character must have zeroes. Given a character χ, if Z_χ is its zero set in X, is it true that $\sigma(Z_\chi) = 0$? Let $Z = \bigcup_n Z_{\chi_n}$; is Z dense in X?

Notice that in case of the hypergroup (J) associated with the Gel'fand pair $(\mathbf{T}\alpha\mathbf{Z}_2, \mathbf{Z}_2)$, the following properties are easily checked: (J) falls into case (A), its characters are real and $\overline{Z} = X$.

2.10. For the next result of this section, we assume the hypergroup to belong to the class (A); $\sup_n \varpi\{\chi_n\} < +\infty$.

Theorem 2.11. *Let $\phi \in \ell^2(\hat{X}, \varpi)$; then for almost every $t \in [0,1]$, the function $\exp\{\mu|f_t|^2\} \in L^1(X, \sigma)$ for $\mu \in \mathbf{R}$. In particular $f_t \in L^p(X, \sigma)$ for any $p < \infty$, for almost every $t \in [0,1]$.*

Proof. If the hypergroup belongs to class (A), then

$$\sum_n |\phi(\chi_n)|^2 |\chi_n(x)|^2 \varpi^2\{\chi_n\} \le M\|\phi\|_2^2 \text{ for every } x \in X.$$

So $\sum_n |\phi(\chi_n)|^2 |\chi_n(x)|^2 \varpi(\chi_n) \le M \sum_n |\phi(\chi_n)|^2 \varpi^2\{\chi_n\}$ for every $x \in X$; by the classical result of [Z,5.5.1.(ii)] we know that by letting

$$\Phi(t) = \sum_n r_n(t)\phi(\chi_n)\varpi\{\chi_n\}$$

where the right hand series converges almost everywhere in $[0,1]$, one has $\int_0^1 \exp \mu|\Phi(t)|^2 dt \le \sum_{k=0}^\infty \frac{k^k}{k!}(\mu C)^k < +\infty$ where $C = \sum |\phi(\chi_n)|^2 \varpi^2\{\chi_n\}$, for some sufficiently small $\mu > 0$. Therefore, $\int_0^1 \exp \mu|f_t(x)|^2 dt \le \sum_{k=0}^\infty \frac{k^k}{k!}(\mu C)^k < +\infty$ for every $x \in X$, and the same $\mu > 0$. Finally $\int_X \int_0^1 \exp \mu|f_t(x)|^2 dt d\sigma(x) < +\infty$.

This proves the claim for sufficiently small $\mu > 0$ by Fubini's theorem. Now let $S_n(t;x) := \sum_{p\leq n} r_p(t)\phi(\chi_p)\chi_p(x)\varpi\{\chi_p\}$. If μ is now arbitrary, the same result applies to $|f_t(x) - S_n(t,x)|^2$ provided n is large enough (depending on μ) for almost every $t \in [0,1]$.

Finally, using again [Z, 5.5.1 (ii)], we see that

$$|f_t(x)|^2 \leq 2[|f_t(x) - S_n(t,x)|^2 + S_n(t,x)|^2]$$

and fixing n large enough, $S_n(t,.)$ is bounded, so the claim is completely proved along the classical lines. ■

2.11.1. Remarks.

(i) This is about everything that can be extended along classical lines in an easy way. Let $C(X)$ be the space of continuous functions on X; it is of interest to decide whether if E is a Sidon set of the hypergroup X, one always has the property $\mathcal{F}(C(X))_{|E} = \ell^2(E,\varpi)$. The problem whether the condition $\sup_n \varpi\{\chi_n\} < +\infty$ is sufficient, necessary or both will be addressed in the forthcoming [G] in full detail. Related results are to be found in $[V]_3$ as well as in $[V]_4$, [L]. However, these papers do not lead to a conclusive statement relative to the above question.

(ii) Theorem 2.11 can also be inferred from $[F - R]$.

3. Multipliers

In this section we just prove some easy extensions of well known results on a compact commutative group. Their mere interest, if any, is the crucial use of 1.4, which already is fairly deep.

Definition 3.1. Let X be a compact commutative hypergroup. If ϕ is a complex valued function on $\hat{X}$, we say that if F, G are two sets of objects living on X (functions, measures,∞), ϕ is a multiplier of type (F,G) if $\phi.\hat{f} \in \mathcal{F}G$ for any $f \in F$. The set of such multipliers is denoted (F,G) as usual.

Let us first mention the following theorem.

Theorem 3.2.

(i) $(C, L^\infty) = (L^\infty, L^\infty) = \mathcal{F}M$ *where C denotes the space $C(X)$, and M denotes the space $M(X)$.*

(ii) $(M,M) = \mathcal{F}M$.

Proof. The proof follows the classical lines used already in the compact abelian group case. One uses 1.4 in an essential way. The result, along with more general multiplier theorems, already appeared in $[V]_2$. ∎

3.3. Finally we explore the following question. Which complex valued functions ϕ on $\hat{X}$ have the property that the series $\sum_n \phi(\chi_n)\hat{f}(\chi_n)\chi_n\varpi\{\chi_n\}$ has uniformly bounded partial sums for each continuous function f on X?

3.4. Definition and Proposition

Let

$$L_b^\infty := \{g \in L^\infty(X,\sigma); \quad \sup_N \| \sum_{n\leq N} \hat{g}(\chi_n)\chi_n\varpi\{\chi_n\}\|_\infty < +\infty\}.$$

We let $\|g\|_{\infty,\sim} = \sup_N \| \sum_{n\leq N} \hat{g}(\chi_n)\chi_n\varpi(\chi_n)\|_\infty$ *for* $g \in L_b^\infty$. *Then* L_b^∞ *is a Banach space and* $\|g\|_\infty \leq \|g\|_{\infty,\sim}$.

Proof. First of all it is clear that $\|g\|_{\infty,\sim}$ is a norm on L_b^∞.

We prove that $\|g\|_\infty \leq \|g\|_{\infty,\sim}$ for $g \in L_b^\infty$. In fact let $g \in L_b^\infty$ and define $g_N := \sum_{n\leq N} \hat{g}(\chi_n)\chi_n\varpi\{\chi_n\}$. The sequence (g_N) is bounded in $L^\infty(X,\sigma)$ hence w^* relatively compact. If Γ is a w^*-accumulation point of (g_N), it is immediate that $\hat{\Gamma}(\chi_n) = \hat{g}(\chi_n)$ for all n, and hence $\Gamma = g$ in L^∞.

Moreover $\|g\|_\infty = \|\Gamma\|_\infty = \sup_{\|f\|_1\leq 1} |\int \Gamma f d\sigma|$ but $\int_X \Gamma f d\sigma = \lim_k \int g_{N_k} f d\sigma$ so that

$$\left|\int \Gamma f d\sigma\right| \leq \sup_k \|g_{N_k}\| \leq \|g\|_{\infty,\sim} \text{ for any } f \in L^1(X,\sigma), \|f\|_1 \leq 1.$$

Finally $\|g\|_\infty \leq \|g\|, \|g\|_{\infty,\sim}$ as claimed.

Now, if (g_k) is a Cauchy sequence in L_b^∞, (g_k) is also Cauchy in L^∞ by the former inequality. Let g be its limit in L^∞.

Fix $\varepsilon > 0$ and $k_\varepsilon \in \mathbf{N}$ such that for k, $k' > k_\varepsilon$ one has $\|g_k - g_{k'}\|_{\infty,\sim} \leq \varepsilon$. Then, letting $S_N(g) = \sum_{n\leq N} \hat{g}(\chi_n)\chi_n\varpi\{\chi_n\}$ for $g \in L^\infty(X,\sigma)$, one has $\|S_N(g_k)-S_N(g_{k'})\|_\infty \leq \varepsilon$ for any N. Letting k' go to infinity and noticing that $\lim_{k'} S_N(g_{k'}) = g$ for any N, in $L^\infty(X,\sigma)$ one has $\|S_N(g_k) - S_N(g)\|_\infty \leq \varepsilon$ for all N and $k \geq k_\varepsilon$. Hence $g = \lim_{k\to\infty} g_k$ in L_b^∞ and the Proposition is proved. ∎

Theorem 3.5. *The multipliers* (C, L_b^∞) *are precisely those of the form* $\phi = \hat{\mu}$ *where* μ *in* $M(X)$ *satisfies*

$$m = \sup_N \| \sum_{n\leq N} \hat{\mu}(\chi_n)\chi_n\varpi\{\chi_n\}\|_1 < +\infty.$$

Moreover, the function ϕ *has the property that* $\sum_n \phi(\chi_n)\hat{f}(\chi_n)\chi_n\varpi\{\chi_n\}$ *has uniformly bounded partial sums for each continuous function* f *on* X *if and only if* $\phi \in (C, L_b^\infty)$ *and in that case*

(i) the series $\sum_n \phi(\chi_n)\hat{f}(\chi_n)\chi_n\varpi\{\chi_n\}$ *is uniformly convergent for each continuous function* f *and*

(ii) the series $\sum_n \phi(\chi_n)\hat{f}(\chi_n)\chi_n\varpi\{\chi_n\}$ *converges in the* L^p *norm for every* $f \in L^p(X, \sigma)$ *for* $1 \le p < \infty$.

Proof. We follow the lines of ([E], p. 258) where virtually anything can be retained, aside from the essential use of 1.4. See below.

First of all if $\phi = \hat{\mu}$ and $\sup_N \| \sum_{n\le N} \hat{\mu}(\chi_n)\chi_n\varpi\{\chi_n\}\|_1 < +\infty$, then for $f \in (CX)$, one has:

$$\begin{aligned}&\sup_n \| \sum_{n\le N} \hat{\mu}(\chi_n)\hat{f}(\chi_n)\chi_n\varpi\{\chi_n\}\|_\infty \\ \le &\sup_N \| \sum \hat{\mu}(\chi_n)\chi_n\varpi\{\chi_n\}\|_1 \cdot \|f\|_\infty\end{aligned}$$

because $\sum_{n\le N} \hat{\mu}(\chi_n)\hat{f}(\chi_n)\chi_n\varpi\{\chi_n\} = S_N(\mu) * f$ where

$$S_N(\mu) = \sum_{n\le N} \hat{\mu}(\chi_n)\hat{f}(\chi_n)\chi_n\varpi\{\chi_n\}.$$

So $\sup_N \| \sum_{n\le N} \hat{\mu}(\chi_n)\hat{f}(\chi_n)\chi_n\varpi\{\chi_n\}\|_\infty \le m\|f\|_\infty$. Therefore the mapping $f \mapsto \mathcal{F}^{-1}(\phi\hat{f})$ is continuous from C into L_b^∞, or $\phi \in (C, L_b^\infty)$.

Conversely, if $\phi \in (C, L_b^\infty)$, the mapping $f \mapsto \mathcal{F}^{-1}(\phi\hat{f})$ is continuous by the closed graph theorem; let $\widehat{U_\phi(f)} = \phi\hat{f}$. Then $\sup_N \|S_N(U_\phi(f))\|_\infty \le m \cdot \|f\|_\infty$ for some constant m. Let $S_N(\phi) = \sum_{n\le N} \phi(\chi_n)\chi_n\varpi\{\chi_n\}$; then $S_N(U_\phi(f)) = S_N(\phi) * f$ and the former inequality shows that $\sup_N \|S_N(\phi)\|_1 < +\infty$.

Therefore, the sequence $(S_N(\phi))_N$ is bounded in $M(X)$, and hence there exists a w^*-accumulation point $\mu \in M(X)$ which is readily seen to satisfy $\phi = \hat{\mu}$ and $\sup_N \|S_N(\mu)\|_1 < +\infty$.

Now, if $f \in C(X)$, there is $\gamma \in \mathcal{P}(X)$ such that $\|f - \gamma\|_\infty \le \varepsilon$. This follows from 1.4 again. (As $\mathcal{P}(X)$ is not always an algebra for pointwise multiplication, the Stone Weierstrass theorem cannot be used here.)

For any N, we have

$$\|S_N(\mu * f) - S_N(\mu * \gamma)\|_\infty \le m\|f - \gamma\|_\infty$$

where $m = \sup_N \|S_N(\mu)\|_1$. Now, $\mu * \gamma \in \mathcal{P}(X)$ and $S_N(\mu * \gamma) = \mu * \gamma$ for $N \geq N_0(\gamma)$. Therefore $\|S_N(\mu * f) - S_{N'}(\mu * f)\|_\infty \leq \epsilon m$ if $N, N' \geq N_0(\gamma)$. Thus the uniform convergence of the series is proved. The last assertion is trivial, using again 1.4 in order to get $\|f - \gamma\|_p \leq \epsilon$ for $1 \leq p < \infty$ for $f \in L^p(X, \sigma)$. ■

3.6. Remark.

In contrast with 3.2, it is rather unexpected that 3.5 will hold in general.

Acknowledgments. The author thanks gratefully the referee for pointing out various important references, and making some useful remarks concerning redaction and style, as well as for numerous interesting mathematical remarks.

References

[B-H] W.R. BLOOM, H. HEYER, *Harmonic analysis of probability measures on hypergroups*, De Gruyter Studies in Mathematics **20** (1995), De Gruyter; Berlin, New-York.

[E] R.E. EDWARDS, *Fourier Series, A modern Introduction Vol. II, Holt*, Rinehart and Winston, Inc. (1967).

[F] J.J.F. FOURNIER, K.A. ROSS, *Random Fourier series on compact abelian hypergroups*, J. Australian Math. Soc. **37** (1984), 45–81.

[G] M.O. GEBUHRER, *Changing signs in a Fourier series on a compact commutative hypergroup*, in preparation.

[G_1] M.O. GEBUHRER, *Bounded measure algebras: a fixed point theorem approach. In "Applications of hypergroups and related measure algebras" (Providence, RI) (O. Gebuhrer, W.C. Connett, A.L. Schwartz, eds.) A.M.S.*, Contemporary Mathematics **183** (1995), 171–190.

[G_2] M.O. GEBUHRER, *Analyse Harmonique sur les espaces de Gel'fand-Levitan et applications à la théorie des semigroupes de convolution, Thèse de Doctorat d'État* (1989 Université Louis Pasteur Strasbourg).

[G-Sc_1] M.O. GEBUHRER, A.L. SCHWARTZ, *Sidon sets and Riesz sets for some measure algebras on the disk.* To appear in Colloquium Mathematicum.

[G-Sc2]M.O. GEBUHRER, A.L. SCHWARTZ, *Sidon sets on compact commutative hypergroups,* in preparation.

[H] S.HELGASON—, *Groups and Geometric Analysis. (Integral Geometry, Invariant Differential Operators and Spherical functions)*, Academic Press (1984).

[L] LASSER, *Lacunarity with respect to Orthogonal Polynomial Sequences*, Acta. Scien. Math. **47** (1984), 391–403.

[S] R. SPECTOR, *Aperçu de la théorie des hypergroupes. In "Analyse Harmonique sur les groupes de Lie" Séminaire Nancy-Strasbourg*, Lecture Notes in Mathematics **497 Springer** (1973-75), 643–673.

[V1] R. VREM, *Harmonic Analysis on Compact Hypergroups*, Pac. J. Math. **85** (1979), 239–251.

[V2] R. VREM, *Thesis.*

[V3] R. VREM, *Lacunarity on Compact Hypergroups*, Math. Z. **164** (1978), 93–104.

[V4] R. VREM, *Independent Sets and Lacunarity for Hypergroup*, J. Austr. Math. Soc. (Series B) **50** (1991), 171–188.

[Z] A. ZYGMUND, *Trigonometrical series*, Dover Publications S 290 (First edition in (1935; references are w.r.t. edition of 1955)).

Email: gebuhrer@math.u-strasbg.fr; Institut de Recherche Mathématique Avancée, Université Louis Pasteur et C.N.R.S. 7, rue René-Descartes, 67084 Strasbourg Cedex

Disintegration of Measures

Henry Helson

Let (X, B), (Y, C), be measurable spaces and let μ be a measure on the product measure space. A *disintegration* of μ is a representation

$$\mu = \int \nu_y d\gamma(y) \ , \tag{1}$$

where for each y in Y, ν_y is a measure on X. The meaning of this formula is that for each ϕ in some specified class of function on $X \times Y$,

$$\int \phi d\mu = \int \Big(\int \phi(x, y) d\nu_y(x) \Big) d\gamma(y) \ . \tag{2}$$

(The definition is not symmetric in the variables; if μ has such a disintegration, it might or might not have a disintegration

$$\mu = \int \eta_x d\rho(x) \ .) \tag{3}$$

The first theorem of this kind was stated by J.L. Doob, although the statement and proof were incomplete. A complete and complicated discussion of the problem is contained in [6]; the interest for statistics of the problem is shown in [5]. An accessible theorem about disintegration is given in [3, p. 318], and ascribed to Abrahamse and Kriete [1]. Bourbaki has another version [2]. Some countability hypothesis always appears. The purpose of this note is to give a theorem that is quite general and whose proof is simple.

Such a result does not belong to pure measure theory. It seems necessary to assume that X and Y are topological spaces of particular kinds, and then the natural statement is that (2) holds for bounded continuous functions ϕ.

Theorem 1. *Let* $\mathbf{T}$ *be the unit circle,* Y *a compact Hausdorff space, and* μ *a probability measure on the Baire field of* $\mathbf{T} \times Y$. *There is a probability measure* γ *on* Y, *and for almost every* y *in* Y *(with respect to* γ*) a probability measure* ν_y *on* $\mathbf{T}$ *such that (2) holds for all continuous functions* ϕ *on the product space.*

If (2) has been shown to hold for all continuous functions ϕ that are products $\psi(x)\eta(y)$, then the Stone–Weierstrass theorem shows that the formula holds for all continuous ϕ. Furthermore, it is enough to prove (2) when ψ is an exponential e^{nix}. That is, it will suffice to find measures γ and ν_y such that

$$\int\int e^{nix}\eta(y)d\mu(x,y) = \int \hat{\nu}_y(n)\eta(y)d\gamma(y) \tag{4}$$

for all integers n and all continuous functions η on Y.

For each integer n, let ζ_n be the complex measure on Y such that

$$\int \eta d\zeta_n = \int\int e^{nix}\eta(y)d\mu(x,y) \tag{5}$$

for continuous functions η. (The right side is obviously a continuous linear functional of η.) Let γ be a probability measure on Y such that each ζ_n is absolutely continuous with respect to γ, and write $d\zeta_n = \alpha_n d\gamma$, where α_n is a function summable for γ.

Lemma. *For a.e.* $(\gamma)\, y, (\alpha_n(y))$ *is a positive definite sequence.*

Let (c_n) be any complex sequence with finite support. We shall show that

$$\sum c_r\bar{c}_s\alpha_{r-s}(y)d\gamma(y) \tag{6}$$

is a positive measure. Then $\sum c_r\bar{c}_s\alpha_{r-s}(y) \geq 0$ a.e. (γ). The same inequality will be true at once for all sequences (c_n) whose values are rational numbers, except for y in a countable union of null sets. Hence the inequality will be true for all sequences without restriction except on the same null set, and this will prove the lemma.

Let η be any continuous non-negative function on Y. We have

$$\begin{aligned}\int \eta(y)\sum c_r\bar{c}_s\alpha_{r-s}(y)d\gamma(y) &= \int \eta(y)\sum c_r\bar{c}_s d\zeta_{r-s}(y)\\ &= \sum c_r\bar{c}_s\int\int e^{-(r-s)ix}\eta(y)d\mu(x,y) \quad (7)\\ &= \int\int\left|\sum c_n e^{-nix}\right|^2\eta(y)d\mu(x,y) \geq 0\,.\end{aligned}$$

This proves the assertion, and the lemma.

By the theorem of Herglotz, there is for a.e. $(\gamma)\, y$ a positive measure

ν_y on $\mathbf{T}$ such that $\hat{\nu}_y(n) = \alpha_n(y)$ for each n. Then

$$\int\int e^{nix}\eta(y)d\mu(x,y) = \int \eta d\zeta_n = \int \eta\alpha_n d\gamma = \int \eta(y)\hat{\nu}_y(n)d\gamma(y) \,, \tag{8}$$

which is the formula (4) that was to be proved.

The measure γ was chosen to be a probability measure, but the ν_y are only known to be positive. The total mass of ν_y is $\hat{\nu}_y(0) = \alpha_0(y)$. Taking $n = 0$ and η the constant function 1 in (8) shows that $\alpha_0(y)d\gamma(y)$ is also a probability measure, so we replace each ν_y by $\nu_y/\alpha_0(y)$ to obtain a probability measure, and γ by $\alpha_0(y)d\gamma(y)$. This completes the proof.

(Actually γ is the projection of μ of the space Y: For each Baire set of E of Y, $\gamma(E) = \mu(\mathbf{T} \times E)$. It could have been defined this way, but that would have complicated the proof.)

In order to make the methods of harmonic analysis applicable, the first factor $\mathbf{T}$ had to be a group. We shall extend the theorem by using the fact that every uncountable standard Borel space is Borel-isomorphic to $\mathbf{T}$.

Theorem 2. *Let X be a standard Borel space and Y a compact Hausdorff space. Give $X \times Y$ the Borel structure that is the product of the Borel field of X and the Baire field of Y. Then every probability measure μ on $X \times Y$ has a disintegration, and (2) holds for all bounded measurable functions ϕ.*

First we show that formula (2), which has been shown to hold for continuous functions ϕ on $\mathbf{T} \times Y$, is actually true for all bounded Baire functions. We need the techniques of [4, Sections 50, 51]. Let S be the set of all bounded real Baire functions ϕ on $\mathbf{T} \times Y$ for which the inner integral on the right side of (2) is a measurable function of y, and the formula holds. The class S is closed under bounded pointwise convergence, by the dominated convergence theorem. Therefore it contains the indicator function of each compact G_δ set. Furthermore, the sets whose indicator functions are in S form a monotone class. The class S_0 of all finite, disjoint unions of proper differences on compact G_δ sets forms a field [4, Theorem 51.F], and the indicator functions of such sets belong to S. The smallest monotone class containing a field is a σ-field, so S contains the σ-field generated by the compact G_δ sets, which is the Baire field. Hence (2) is true if ϕ is the indicator function of a Baire set, and therefore holds for any bounded Baire function.

If X is a countable set, the theorem can easily be proved directly. Let X be an uncountable standard Borel space, and let θ be a mapping of X onto $\mathbf{T}$ that is a Borel isomorphism [7]. The Baire field $\mathcal{T}$ of $\mathbf{T}\times Y$ is the product of the Borel field of $\mathbf{T}$ and the Baire field of Y. Form the product $\mathcal{X}$ of the Borel field of X with the Baire field of Y. Then the mapping θ induces a Borel isomorphism of $X \times Y$ and $\mathbf{T} \times Y$, and the statement of the theorem is obvious.

If X is a compact *metric* space, its Borel field makes it a standard Borel space, so that Theorem 2 holds. In this case X is separable, but Y need not be.

References

[1] Abrahamse, M.B., and Kriete, T.L., The spectral multiplicity of a multiplication operator, *Indiana Math. J.* **22** (1973), 845–847.

[2] Bourbaki, N., *Éléments de mathématiques*, Livre VI, Hermann et Cie, 1959.

[3] Conway, J.B., *The Theory of Subnormal Operators*, Amer. Math. Soc., 1991.

[4] Halmos, P.R., *Measure Theory*, Springer-Verlag, 1974.

[5] Le Cam, L., *Asymptotic Methods in Statistical Decision Theory*, Springer-Verlag, 1986.

[6] Pachl, J.K., Disintegration and compact measures, *Math. Scand.* **43** (1978), 157–168.

[7] Parthasarathy, K.R., *Probability Measures on Metric Spaces*, Academic Press, 1967.

Department of Mathematics, University of California, Berkeley

Multipliers of de Branges–Rovnyak spaces II

Benjamin A. Lotto and Donald Sarason*

Abstract

Given a nonextreme point b of the unit ball of H^∞, the multipliers of the de Branges-Rovnyak space $\mathcal{H}(b)$ lie in an auxiliary space $\mathcal{M}(\bar{a}) \cap H^\infty$, where a is a function in H^∞ that is associated with b and $\mathcal{M}(\bar{a})$ is the range of the Toeplitz operator $T_{\bar{a}}$ on H^2. An example is constructed here to show that $\mathcal{M}(\bar{a}) \cap H^\infty$ need not be an algebra. This contrasts with the case where b is an extreme point, where the analogous auxiliary space is always an algebra.

For $1 \leq p \leq \infty$, let L^p denote the usual Lebesgue space of functions on the unit circle $\partial\mathbf{D}$ and let H^p denote the classical Hardy space, thought of either as a subspace of L^p or as a space of holomorphic functions on the unit disk $\mathbf{D}$. For ϕ in L^∞, the *Toeplitz operator with symbol* ϕ is the operator T_ϕ defined by the formula $T_\phi f = P_+(\phi f)$, where P_+ denotes the orthogonal projection from L^2 onto H^2.

Given a function b in the unit ball of H^∞, we define $\mathcal{H}(b)$, the *de Branges-Rovnyak space with symbol* b, to be the range of the operator $(1 - T_b T_{\bar{b}})^{1/2}$ on H^2. We give $\mathcal{H}(b)$ the range norm

$$\left\|(1 - T_b T_{\bar{b}})^{1/2} f\right\|_{\mathcal{H}(b)} = \|f\|_2$$

where $f \perp \ker(1 - T_b T_{\bar{b}})^{1/2}$. This norm makes $\mathcal{H}(b)$ into a Hilbert space. The basic properties of de Branges-Rovnyak spaces can be found in [6].

We are concerned with the multipliers of $\mathcal{H}(b)$, that is, the functions m in H^∞ for which mf belongs to $\mathcal{H}(b)$ whenever f does. Multipliers of $\mathcal{H}(b)$ are studied in [3, 4]. The theory bifurcates according to whether b is or is not an extreme point of the unit ball of H^∞. In the extreme point case [3], the multipliers of $\mathcal{H}(b)$ belong to an auxiliary space called $K^\infty(\rho)$, which turns out to be an algebra of bounded holomorphic functions on $\mathbf{D}$. It is the purpose of this note to show that the situation is different in the nonextreme point case.

*First author partially supported by NSF grant DMS-9502983

When b is in the unit ball of H^∞ but not an extreme point, a theorem of K. de Leeuw and W. Rudin [1] tells us that $\log(1-|b|^2)$ is integrable over $\partial\mathbf{D}$, so there is an H^∞ function a such that $|a|^2 = 1-|b|^2$ on $\mathbf{D}$ [5, Theorem 17.16, p. 343]. We take a to be outer and positive at the origin; this defines a uniquely. The space $\mathcal{M}(\bar{a})$ is defined to be the range of $T_{\bar{a}}$ on H^2; as with $\mathcal{H}(b)$, we make $\mathcal{M}(\bar{a})$ into a Hilbert space by giving it the range norm

$$\|T_{\bar{a}}f\|_{\mathcal{M}(\bar{a})} = \|f\|_2.$$

(The kernel of $T_{\bar{a}}$ is trivial so we don't need to include the extra condition that $f \perp \ker T_{\bar{a}}$.) It turns out that every multiplier of $\mathcal{H}(b)$ belongs to $\mathcal{M}(\bar{a})$ and that the space $\mathcal{M}(\bar{a}) \cap H^\infty$ is the auxiliary space analogous to the space $K^\infty(\rho)$ mentioned above. The question thus arises whether $\mathcal{M}(\bar{a}) \cap H^\infty$ is an algebra. We shall show that it need not be. Specifically, we construct a nonextreme point b such that b itself belongs to $\mathcal{M}(\bar{a})$ but b^2 does not. The property that we will use to identify whether or not functions belong to $\mathcal{M}(\bar{a})$ is given in the following definition.

Definition 1. *Suppose that f and g belong to H^∞. We say that (f,g) is an H^2-corona pair if there exist functions ϕ and ψ belonging to H^2 with $f\phi + g\psi = 1$.*

K. C. Lin [2] has found both necessary and sufficient conditions on f and g for the pair (f,g) to be an H^2-corona pair. These conditions involve an estimate of the nontangential maximal function of $1/(|f|^2+|g|^2)^{1/2}$. Unfortunately, there is a gap between the two conditions. It would be nice to have simple necessary and sufficient conditions on f and g for (f,g) to be an H^2-corona pair.

Our first result connects being an H^2-corona pair with membership in $\mathcal{M}(\bar{a})$. We write H_0^2 for the space of functions in H^2 that vanish at the origin and $\overline{H^2}$ (respectively $\overline{H_0^2}$) for the space of complex conjugates of functions in H^2 (respectively H_0^2). Note that $(H^2)^\perp = \overline{H_0^2}$.

Theorem 1. *Suppose that f is in H^∞ and that $(c-|f|^2)/|a|$ belongs to L^∞ for some positive constant c. Then f belongs to $\mathcal{M}(\bar{a})$ if and only if (f,a) is an H^2-corona pair.*

Proof. First, suppose that f belongs to $\mathcal{M}(\bar{a})$ and write $f = T_{\bar{a}}g$ with $g \in H^2$. Then $f - \bar{a}g$ belongs to $\overline{H_0^2}$ and so $\bar{f} - a\bar{g}$ belongs to H_0^2. Denote this function by ϕ. Multiplying by f keeps us in H_0^2, so we find that

$$|f|^2 - af\bar{g} = c - a\left[\frac{c-|f|^2}{a} + f\bar{g}\right]$$

belongs to H_0^2. Hence the second term on the right belongs to H^2. Since a is outer and the quantity in brackets belongs to L^2, we can divide by a and conclude that the quantity in brackets actually belongs to H^2. Denote this quantity by ψ. Then ϕ and ψ belong to H^2 and $f\phi + a\psi = |f|^2 - af\bar{g} + (c - |f|^2) + af\bar{g} = c$. Dividing ϕ and ψ by c, we see that (f, a) is an H^2-corona pair.

For the other direction, suppose that (f, a) is an H^2-corona pair. Let ϕ and ψ be H^2 functions that satisfy $f\phi + a\psi = 1$. Then $\bar{f}\bar{\phi} + \bar{a}\bar{\psi} = 1$, so

$$\begin{aligned} f &= |f|^2\bar{\phi} + \bar{a}f\bar{\psi} \\ &= \bar{a}\left[-\frac{c - |f|^2}{\bar{a}}\bar{\phi} + f\bar{\psi}\right] + c\bar{\phi}. \end{aligned}$$

Let ρ denote the term in brackets. Then ρ belongs to L^2. Applying P_+ to the equality above we get

$$f = P_+(\bar{a}\rho) + c\overline{\phi(0)} = T_{\bar{a}}\left(P_+\rho + c\overline{\phi(0)}\big/\overline{a(0)}\right),$$

so f belongs to $\mathcal{M}(\bar{a})$ as desired. ∎

Remark 1. The first part of the proof goes through as written under the weaker hypothesis that $(c - |f|^2)/|a| \in L^2$.

Remark 2. The second part of the proof goes through essentially as written under the weaker hypothesis that $(c - |f|^2)/|a| \in L^2$ if we also assume the stronger hypothesis that (f, a) is a corona pair. (This means that (f, a) satisfies the conclusion of Carleson's Corona Theorem, that is, there exist functions ϕ and ψ in H^∞ such that $f\phi + a\psi = 1$.)

Remark 3. If we keep the hypothesis that $(c - |f|^2)/|a| \in L^\infty$ and assume that (f, a) is a corona pair, we can obtain the stronger conclusion that f is actually a multiplier of $\mathcal{M}(\bar{a})$. To see this, let g be in $\mathcal{M}(\bar{a})$ and write $g = T_{\bar{a}}h$ for some h in H^2. Arguing as above, we find that

$$\begin{aligned} fg &= |f|^2 g\bar{\phi} + \bar{a}fg\bar{\psi} \\ &= \bar{a}\left[ch\bar{\phi} + \frac{c - |f|^2}{\bar{a}}g\bar{\phi} + fg\bar{\psi}\right] + c(g - \bar{a}h)\bar{\phi}. \end{aligned}$$

The second term on the right belongs to $\overline{H_0^2}$ and the term in brackets (which we'll call ρ as above) is in L^2. It follows as before that $fg = T_{\bar{a}}(P_+\rho)$, so that fg belongs to $\mathcal{M}(\bar{a})$.

Our next result is a necessary condition for a pair of functions to be an H^2-corona pair.

Lemma 1. *Suppose that (f, g) is an H^2-corona pair. Then there is a constant $C > 0$ such that*

$$|f(z)|^2 + |g(z)|^2 \geq C(1 - |z|^2)$$

for all z in $\mathbf{D}$.

Proof. Clearly (f, g) is an H^2-corona pair if and only if the function 1 is in the range of the operator $T(\phi \oplus \psi) = f\phi + g\psi$ on $H^2 \oplus H^2$. By a well-known condition, this happens if and only if there is a constant $C > 0$ such that

$$\left|\langle h, 1\rangle\right|^2 \leq C\|T^*h\|^2$$

for all h in H^2. Now $T^*h = T_{\bar{f}}h \oplus T_{\bar{g}}h$, so taking $h = k_z$, the reproducing kernel function for H^2 at z in $\mathbf{D}$, we obtain

$$1 = \left|\langle k_z, 1\rangle\right|^2 \leq C\left(\|T_{\bar{f}}k_z\|^2 + \|T_{\bar{g}}k_z\|^2\right) = C\frac{|f(z)|^2 + |g(z)|^2}{1 - |z|^2},$$

as desired. ■

Remark 4. We may replace $1 - |z|^2$ by $1 - |z|$ in the above result because the ratio of the two terms is bounded above and below on $\mathbf{D}$.

We are now ready to construct our example. Fix $\alpha, \beta > 0$ (to be determined later) and define the function $w(\theta)$ on $\partial\mathbf{D}$ by

$$w(\theta) = \begin{cases} \theta^\alpha & \text{if } 0 < \theta < 1/2 \\ \left(1 - |\theta|^{2\beta}\right)^{1/2} & \text{if } -1/2 < \theta < 0; \\ 1/\sqrt{2} & \text{otherwise.} \end{cases}$$

Clearly w belongs to L^∞ and $\log w$ is integrable. We therefore may define b to be the outer function in H^∞ with $|b(\theta)| = w(\theta)$. It is easy to check that $\log(1 - |b|^2)$ is integrable, so by the previously mentioned theorem of de Leeuw and Rudin, b is not an extreme point of the unit ball of H^∞. The corresponding function a satisfies

$$|a(\theta)| = \begin{cases} \left(1 - \theta^{2\alpha}\right)^{1/2} & \text{if } 0 < \theta < 1/2 \\ |\theta|^\beta & \text{if } -1/2 < \theta < 0; \\ 1/\sqrt{2} & \text{otherwise.} \end{cases}$$

We need to make an estimate on b.

Lemma 2. *For every $\epsilon > 0$, $|b(r)|/(1 - r)^{\frac{\alpha}{2} - \epsilon} \to 0$ as $r \to 1^-$.*

Proof. Let $P_z(\theta)$ denote the Poisson kernel for the point z in $\mathbf{D}$. For $\theta_0 < 1/2$ and $0 < r < 1$ we have

$$\begin{aligned}\log|b(r)| &= \frac{1}{2\pi}\int_0^{2\pi} P_r(\theta)\log|b(\theta)|\,d\theta \\ &\leq \frac{1}{2\pi}\int_0^{\theta_0} P_r(\theta)\log\theta^\alpha\,d\theta \\ &\leq \frac{\alpha\log\theta_0}{2\pi}\int_0^{\theta_0} P_r(\theta)\,d\theta.\end{aligned}$$

Substituting in

$$P_r(\theta) = \frac{1-r^2}{1+r^2-2r\cos(\theta)}$$

and integrating gives

$$\log|b(r)| \leq \frac{\alpha\log\theta_0}{\pi}\arctan\left(\frac{1+r}{1-r}\tan\frac{\theta_0}{2}\right).$$

Fix $\delta > 0$ and take $\theta_0 = (1-r)^{1-\delta}$. As $r \to 1^-$, then, we have

$$\arctan\left(\frac{1+r}{1-r}\tan\frac{\theta_0}{2}\right) \to \frac{\pi}{2}.$$

Taking r close enough to 1 gives us

$$\begin{aligned}\log|b(r)| &\leq \frac{\alpha\log(1-r)^{1-\delta}}{\pi}\left(\frac{\pi}{2}-\delta\right) \\ &= (1-\delta)\left(1-\frac{2\delta}{\pi}\right)\frac{\alpha}{2}\log(1-r) \\ &\leq \left(\frac{\alpha}{2}-\epsilon\right)\log(1-r)\,,\end{aligned}$$

when δ is chosen small enough so that $(1-\delta)(1-2\delta/\pi)\alpha/2 \geq \alpha/2 - \epsilon$. Hence $|b(r)|/(1-r)^{\frac{\alpha}{2}-\epsilon}$ is bounded above. But ϵ was arbitrary, so $|b(r)|/(1-r)^{\frac{\alpha}{2}-\epsilon'}$ is bounded above for a slightly smaller ϵ'. This implies that $|b(r)|/(1-r)^{\frac{\alpha}{2}-\epsilon}$ actually tends to 0 as $r \to 1^-$. ∎

Remark 5. The same estimate applies to a and gives $|a(r)|/(1-r)^{\frac{\beta}{2}-\epsilon} \to 0$ as $r \to 1^-$ for any $\epsilon > 0$.

Remark 6. Similar reasoning shows that for any $\epsilon > 0$, $|b(z)|/(1-|z|)^{\alpha-\epsilon} \to 0$ as $z \to 1$ along some line segment in $\mathbf{D}$ that terminates at 1. Let ϕ be the angle that the line makes with a vertical line, measured

1. Let ϕ be the angle that the line makes with a vertical line, measured counterclockwise (so the case of the previous lemma is $\phi = \pi/2$). Write $z = re^{i\rho}$ for the point on the segment at distance r from the origin. Following the reasoning in the proof of the lemma, we find that

$$\begin{aligned}\log|b(re^{i\rho})| &\leq \frac{\alpha \log \theta_0}{2\pi} \int_0^{\theta_0} P_z(\theta)\, d\theta \\ &= \frac{\alpha \log \theta_0}{\pi} \left[\arctan \left(\frac{1+r}{1-r} \tan(\theta) \right) \right]_{\theta=-\rho}^{\theta=\theta_0-\rho}.\end{aligned}$$

Take $r \to 1$ and let $\theta_0 = (1-r)^{1-\delta}$ as before. The estimate $\rho \approx (1-r)/\tan\phi$ gives that the arctangent term tends to $\pi/2 + \arctan(2\cot\phi)$, so we find that

$$\log|b(z)| \leq (1-r)^{c\alpha-\epsilon}$$

where $c = 1/2 + \arctan(2\cot\phi)/\pi$.

Theorem 2. *Let b and a be as above.*

1. *If $f = b^n$, then $\dfrac{1-|f|^2}{|a|}$ is in L^∞.*
2. *If $\alpha < 1/2$, then $b \in \mathcal{M}(\bar{a})$.*
3. *If $n\alpha > 1$ and $\beta > 1$, then $b^n \notin \mathcal{M}(\bar{a})$.*
4. *If $\alpha < 1/2$, $n\alpha > 1$, and $\beta > 1$, then $\mathcal{M}(\bar{a}) \cap H^\infty$ is not an algebra.*

Proof. For part 1, note that

$$\frac{1-|b^n|^2}{|a|} = \frac{1-|b|^{2n}}{|a|} = |a|(1 + |b|^2 + \cdots + |b|^{2n-1})$$

since $1 - |b|^2 = |a|^2$. This belongs to L^∞ since both a and b do.

For part 2, note that if $\alpha < 1/2$, then $1/|b|$ is in L^2. Since b is outer, it follows that $1/b$ is in H^2. Hence (b, a) is an H^2 corona pair via the identity $b \cdot (1/b) + a \cdot 0 = 1$. It follows from part 1 and Theorem 1 that b is in $\mathcal{M}(\bar{a})$.

Now part 3. It follows from Lemma 2 that $|b^n(r)|^2/(1-r) \to 0$ as $r \to 1^-$. Similarly, from the first remark following Lemma 2, $a(r)|^2/(1-r) \to 0$ as $r \to 1^-$. It now follows from Lemma 1 that (b^n, a) is not a corona pair. Hence from part 1 and Theorem 1, b^n does not belong to $\mathcal{M}(\bar{a})$.

Finally, part 4 follows immediately from parts 2 and 3. ■

Remark 7. The hypotheses for part 4 of Theorem 2 can be achieved by taking α to be slightly less than $1/2$, $n = 3$, and β to be anything larger than 1. This gives an example of a function b for which b is in $\mathcal{M}(\bar{a})$ but b^3 is not in $\mathcal{M}(\bar{a})$.

Remark 8. In order to get the promised example (b in $\mathcal{M}(\bar{a})$ but b^2 not in $\mathcal{M}(\bar{a})$), we have to modify the previous theorem slightly, making the estimates of b and a along a segment in **D** that terminates at 1 different from the radius. We use the estimate from Remark as well as the notation introduced there. Let $z = re^{i\rho} \to 1$ along a very nearly vertical segment. Then $|b(re^{i\rho})|/(1-r)^{\alpha-\epsilon} \to 0$ as $r \to 1^-$. If we take α to be less than $1/2$, we find that $1/b$ is in H^2 as above so that (b, a) is an H^2-corona pair and so b is in $\mathcal{M}(\bar{a})$. Now $|b^2(re^{i\rho})|^2/(1-r)^{4(\alpha-\epsilon)} \to 0$ as $r \to 1^-$. Choose $\alpha > 1/4$ and $\epsilon > 0$ small enough so that $4(\alpha - \epsilon) > 1$. (These choices will dictate the exact segment along which we make our estimate.) This being done, take β large enough so that $|a(re^{i\rho})|^2/(1-r) \to 0$ as $r \to 1^-$. We can now show as above that (b^2, a) is not an H^2-corona pair, so that b^2 is not in $\mathcal{M}(\bar{a})$.

As a final remark, we note we have a factor of two to play with in this example (we can choose $1/4 < \alpha < 1/2$). The gap between Lin's necessary and sufficient conditions for a pair of functions to be an H^2-corona pair also involves a factor of two. We are curious if there is any significance to this.

Acknowledgements. The first author would like to thank Dinesh Singh and the University of Delhi for their hospitality during the Second International Conference on Harmonic Analysis.

References

[1] K. de Leeuw and W. Rudin, Extreme point and extremum problems in H^1, *Pacific J. Math.*, **8** (1958), 467–485.

[2] Kai-Ching Lin, On the H^p solutions to the corona problem, *Bull. Sci. Math.*, **118** (1994), 271–286.

[3] B. A. Lotto and D. Sarason, Multiplicative structure of de Branges's spaces, *Revista Matematica Iberoamericana*, **7** (1991), 183–220.

[4] B. A. Lotto and D. Sarason, Multipliers of de Branges-Rovnyak spaces, *Indiana Univ. Math. J.*, **42** (1993), 907–920.

[5] W. Rudin, *Real And Complex Analysis*, McGraw-Hill, New York, third edition, 1987.

[6] D. Sarason, *Sub-Hardy Spaces in the Unit Disk*, volume 10 of *The University of Arkansas Lecture Notes in the Mathematical Sciences*, Wiley, New York, 1995.

Email: belotto@vassar.edu; Department of Mathematics, Vassar College, 124 Raymond Avenue, Poughkeepsie, NY 12601

Email: sarason@math.berkeley.edu, Department of Mathematics, University of California, Berkeley, CA 94720

On Hartman Uniform Distribution and Measures on Compact Spaces

R. Nair

Abstract

Given a sequence of natural numbers $k = (k_n)_{n=1}^{\infty}$, we say it is Hartman uniformly distributed if

$$\lim_{N\to\infty} \frac{1}{N}\sum_{n=1}^{N} e^{2\pi i k_n x} = 0,$$

for every non-integer real number x. This property of k being Hartman uniform distributed is interesting in subsequence ergodic theory because if for some p in $[1,2]$, for a function $f \in L^p(X,\beta,\mu)$ we have the limit

$$\lim_{N\to\infty} \frac{1}{N}\sum_{n=1}^{N} f(T^{k_n}x) = \ell_{T,f}(x)$$

existing almost everywhere with respect to the measure μ, then just the ergodicity of the dynamical system (X,β,μ,T) implies that $\ell_{T,f}(x) = \int_X f d\mu$, which is of course useful with regard to applications. Not every sequence for which a pointwise convergence theorem holds has this property. In this paper we use this observation to give some results about invariant measures for continuous maps of compact metric spaces.

1. Introduction

Let (X,β,μ) be a probability space and let $T : X \to X$ be a measurable map that is also measure preserving; that is, given $A \in \beta$, we have $\mu(T^{-1}A) = \mu(A)$, where $T^{-1}A$ denotes the set $\{x \in X : Tx \in A\}$. In this paper we say a sequence $k = (k_i)_{i=1}^{\infty}$ is L^p good universal, if for each probability space (X,β,μ), each measurable measure preserving map T of it and for all functions $f \in L^p(X,\beta,\mu)$

we have the limit

$$\lim_{N\to\infty}\frac{1}{N}\sum_{i=1}^{N} f(T^{k_i}x) \;=\; \ell_{T,f}(x),$$

existing almost everywhere with respect to the measure μ. A number of sub-sequences of the natural numbers have been shown to be L^p good universal in the last few years. For instance if $\phi : \mathbb{N} \to \mathbb{N}$ is a polynomial mapping the integers to themselves and $(p_n)_{n=1}^{\infty}$ is the sequence of rational primes, then the the sequences of integers $(\phi(n))_{n=1}^{\infty}$ and $(\phi(p_n))_{n=1}^{\infty}$ are L^p good universal for p in $(1,\infty]$. For the detailed proofs see [Bo2], [Na1] and [Na2] respectively. For a set S let χ_S denote its characteristic function. Suppose $K = (n_k)_{k=1}^{\infty} \subseteq \mathbb{N}$ is a strictly increasing sequence of natural numbers. By identifying K with its characteristic function χ_K we may view it as a point in $\Lambda = \{0,1\}^{\mathbb{N}}$, the set of maps from $\mathbb{N}$ to $\{0,1\}$. Conversely we can also view Λ as the space of strictly increasing sequences of natural numbers like K. We endow Λ with a probability measure by viewing it as the Cartesian product $\Lambda = \prod_{n=1}^{\infty} X_n$ where for each n ($n = 1,2,\ldots$) we have $X_n = \{0,1\}$ and specify the probability π_n on X_n by $\pi_n(\{1\}) = q_n$ with $0 \le q_n \le 1$ and $\pi_n(\{0\}) = 1 - q_n$ where the sequence $(q_n)_{n=1}^{\infty}$ satisfies $\lim_{n\to\infty} q_n n(\log\log n)^{-1} = 0$. The desired probability measure on Λ is the corresponding product measure $\pi = \prod_{n=1}^{\infty}\pi_n$. The underlying σ-algebra β is that generated by the "cylinders"

$$\{\lambda = (\lambda_n)_{n=1}^{\infty} \in \Lambda : \lambda_{i_1} = \alpha_{i_1}, \ldots, \lambda_{i_r} = \alpha_{i_r}\}$$

for all possible choices of $i_1,\ldots,i_r$ and $\alpha_{i_1},\ldots,\alpha_{i_r}$. Then almost every K in (Λ,β,π) is also L^p good universal for each p in $(1,\infty]$ [Bo1]. Suppose for a real number y we denote its integer part by $[y]$. Suppose $g(x)$ is a function from $[1,\infty)$ to itself whose derivative increases to infinity with its argument. Let k_n denote $[g(n)]$ ($n = 1,2,\ldots$) and let A_M denote the cardinality of $\{n : k_n \le M\}$. Suppose for a function $a : [1,\infty) \to [1,\infty)$ increasing to infinity as its argument does, that we set

$$b(M) = \sup_{\langle\alpha\rangle\in[\frac{1}{a(M)},\frac{1}{2})} \left|\sum_{n:k_n\le M} e^{2\pi i\alpha k_n}\right|.$$

Suppose also that for some decreasing function $c : [1,\infty) \to (0,\infty)$

there exists $C > 0$ such that

$$\frac{b(M) + A_{[a(M)]} + \frac{M}{a(M)}}{A_M} \leq Cc(M),$$

and we have

$$\sum_{s=1}^{\infty} c(\theta^s) < \infty.$$

The sequences $k = (k_n)_{n=1}^{\infty}$ we have just described are also L^p good universal for each p in $(1, \infty]$. Sequences that satisfy this condition include the cases where $g(n) = n^\beta$ where β is non-integer and greater than one; $g(n) = P(n)$, where $P(x) = \alpha_0 + \alpha_1 x + \cdots + \alpha_k x^k$ with $\alpha_1, \ldots, \alpha_k$ are not all rational multiples of the same real number and $g(n) = e^{(\log n)^t}$ with $t \in (1, \frac{3}{2})$ [Na4].

We say a sequence is Hartman uniformly distributed if for x not an integer

$$\lim_{N \to \infty} \frac{1}{N} \sum_{n=1}^{N} e^{2\pi i k_n x} = 0.$$

A sequence being Hartman uniformly distributed has a couple of equivalent formulations. Firstly, a sequence is Hartman uniformly distributed if and only if it is uniformly distributed on the Bohr compactification of the integers. Secondly, a sequence is Hartman uniformly distributed if and only if for each irrational number α, if $\langle y \rangle$ denotes the fractional part of y, the sequence $(\langle k_n \alpha \rangle)_{n=1}^{\infty}$ is uniformly distributed modulo one and further the sequence k itself is uniformly distributed among the residue classes modulo m for every natural number m greater than or equal to two. For more on Hartman uniform distribution see [H] or [KN]. See also [N] for more on uniformly distributed sequences of integers and [Na3] for a large class of sequences of integers which are Hartman uniformly distributed, not all of which however are known to be L^p good universal. For example, the sequences $(\phi(n))_{n=1}^{\infty}$ and $(\phi(p_n))_{n=1}^{\infty}$, mentioned earlier, are not in general Hartman uniformly distributed but the other examples mentioned above are. Recall that we say (X, β, μ, T) is ergodic if $T^{-1}A = A$ for any $A \in \beta$ implies $\mu(A)$ is either zero or one. Hartman uniformly distributed sequences are interesting in subsequence ergodic theory because for them ergodicity of the underlying dynamical system (X, β, μ, T) implies that the limit of the ergodic average is equal to the the integral of the function almost everywhere. This is of course very useful with regard to applications. This is not the case for all sequences (cf. [AN]). In Section 2 we describe some new formulations of ergodicity. In Section 3 we apply these to study weak convergence of measures on compact spaces and generalize to good universal Hartman uniformly distributed sequences results previously only known for $k_n = n$ ($n = 1, 2, \ldots$).

2. Conditions equivalent to ergodicity

We say a sequence of complex numbers $(a_n)_{n=-\infty}^{\infty}$ is positive definite if for any sequence $(z_n)_{n=-\infty}^{\infty}$ of complex numbers only a finite number of whose terms are non-zero

$$\sum_{n,m} a_{n-m} z_n \overline{z_m} \geq 0.$$

We have the following lemma due to Herglotz [Ka p. 38].

Lemma 2.1. *A sequence of complex numbers is positive definite if and only if there exists a finite measure ω on $\mathbb{T} = \mathbb{R}/\mathbb{Z}$ such that*

$$a_n = \int_{\mathbb{T}} z^n d\omega(z). \qquad (n \in \mathbb{Z})$$

Let $U = U_T$ denote the Koopman operator defined pointwise on $L^p(X, \beta, \mu)$ for $1 \leq p \leq \infty$ by $U(f(x)) = f(Tx)$. Recall that because T is measure preserving, for each function f in $L^1(X, \beta, \mu)$ we have $\int_X f(Tx) d\mu = \int_X f(x) d\mu$ and so in particular $||Uf||_2 = ||f||_2$ where as usual for $1 \leq p \leq \infty$, the notation $||.||_p$ refers to the L^p norm. This means that U is an isometry on the Hilbert space L^2. Let $\langle .,. \rangle$ denote the standard inner product on $L^2(X, \beta, \mu)$. If we denote the adjoint of U by U^{-1}, then the sequence $(\langle U^n f, f \rangle)_{n=-\infty}^{\infty}$ is positive definite; hence by Lemma 2.1 there exists a measure ω_f satisfying

$$\langle U^n f, f \rangle = \int_T z^n d\omega_f(z). \qquad (n \in Z)$$

The measure ω_f is called the spectral measure.

Throughout this section we assume that $k = (k_n)_{n=1}^{\infty}$ is L^p good universal for a fixed $p \in [1, 2]$. We have the following lemma.

Lemma 2.2. *If for p in $[1,2]$ and $k = (k_n)_{n=1}^{\infty}$ which is Hartman uniformly distributed we have*

$$\lim_{N\to\infty} \frac{1}{N}\sum_{n=1}^{N} f(T^{k_n}x) = \ell_{T,f}(x)$$

in L^p norm, then $\ell_{T,f}(Tx) = \ell_{T,f}(x)$ μ almost everywhere. In consequence, if (X,β,μ,T) is ergodic then $\ell_{T,f}(x) = \int_X f d\mu$ μ almost everywhere.

Proof. First assume that $p = 2$. Then

$$\left\|\frac{1}{N}\sum_{n=1}^{N} f(T^{k_n+1}x) - \frac{1}{N}\sum_{n=1}^{N} f(T^{k_n}x)\right\|_2^2$$

$$= \frac{1}{N^2}\sum_{0<k_i,k_j\le N}\int_X (f\circ T^{k_i+1} - f\circ T^{k_i}) \times (f\circ T^{k_j+1} - f\circ T^{k_j})(x)d\mu(x)$$

$$= \frac{1}{N^2}\sum_{0<k_i,k_j\le N}\Big(\langle U^{k_i+1-(k_j+1)}f,f\rangle - \langle U^{k_i+1-k_j}f,f\rangle - \langle U^{k_i-(k_j+1)}f,f\rangle + \langle U^{k_i-k_j}f,f\rangle\Big)$$

$$= \frac{1}{N^2}\sum_{0<k_i,k_j\le N}\int_T (2-z-z^{-1})z^{k_i-k_j}d\omega(z)$$

$$= 4\int_{\mathbb{T}} \sin^2\frac{\theta}{2}\left|\frac{1}{N}\sum_{n=1}^{N} z^{k_n}\right|^2 d\omega_f(z) ,$$

where $z = e^{i\theta}$. This tends to zero with N because $k = (k_n)_{n=1}^{\infty}$ is Hartman uniformly distributed. The Lemma follows for general p in $(1,2]$ as L^2 is dense in L^p. ∎

We now come to the following central result, which together with its proof in the case where $k_n = n$ $(n = 1,2,\ldots)$ is already known [W].

Theorem 2.3. *Consider the following statements:*

(i) the dynamical system (X,β,μ,T) is such that for each pair of

elements A and B in β we have

$$\lim_{N\to\infty}\frac{1}{N}\sum_{n=1}^{N}\mu(A\cap T^{-k_n}B) \;=\; \mu(A)\mu(B);$$

(ii) for each pair f, g of functions in $L^2(X,\beta,\mu)$ we have

$$\lim_{N\to\infty}\frac{1}{N}\sum_{i=1}^{N}\langle U^{k_i}f,g\rangle \;=\; \langle f,1\rangle\langle 1,g\rangle;$$

(iii) if f is in $L^2(X,\beta,\mu)$, $\int_X fd\mu = 0$ and ω_f is the spectral measure defined above by the sequence $(\langle U^n f, f\rangle)_{n\in\mathbb{Z}}$, then

$$\lim_{N\to\infty}\int_{\mathbb{T}}\left|\frac{1}{N}\sum_{i=1}^{N}z^{k_i}\right|^2 d\omega_f(z) = 0;$$

(iv) if f is in $L^p(X,\beta,\mu)$ for p in $[1,2]$ then

$$\lim_{N\to\infty}\frac{1}{N}\sum_{i=1}^{N}f(T^{k_i}x) \;=\; \int_X fd\mu,$$

in $L^p(X,\beta,\mu)$ norm;

and

(v) if f is in $L^p(X,\beta,\mu)$ for p in $[1,2]$ then

$$\lim_{N\to\infty}\frac{1}{N}\sum_{i=1}^{N}f(T^{k_i}x) \;=\; \int_X fd\mu,$$

almost everywhere with respect to μ.

(vi) T is ergodic.

Then, irrespective of whether $(k_n)_{n=1}^{\infty}$ is Hartman uniformly distributed or not, (i) is equivalent to (ii), and (iii) is equivalent to (iv). Also (ii) implies (vi), and (v) implies (i). Further in light of Lemma 2.2 (vi) implies (iv) and (v), so assuming $(k_n)_{n=1}^{\infty}$ is Hartman uniform distributed we see that all the conditions are equivalent.

Proof. That Theorem 2.3 (ii) implies Theorem 2.3 (i) is manifest because (i) is the special case of (ii) where $f = \chi_A$ and $g = \chi_B$.

Proof that Theorem 2.3 (ii) implies Theorem 2.3 (i). In the special case where for two sets A and B in β we set $a = \chi_A$ and $b = \chi_B$ (i) may be rewritten

$$\lim_{N\to\infty} \frac{1}{N}\sum_{i=1}^{N}\langle U^{k_i}a, b\rangle = \langle a, 1\rangle\langle 1, b\rangle. \tag{2.1}$$

Also, by taking linear combinations of characteristic functions, this statement is also seen to remain true for simple functions a and b. For two arbitrary $L^2(X, \beta, \mu)$ functions f and g given $\epsilon > 0$ we can find simple functions a and b such that $||f - a||_2 \leq \epsilon$ and $||g - b||_2 \leq \epsilon$. Further by (2.1) we can find a possibly large natural number $n = n(\epsilon)$ such that if $N \geq n(\epsilon)$ then

$$\left|\frac{1}{N}\sum_{i=1}^{N}\langle U^{k_i}a, b\rangle - \langle a, 1\rangle\langle 1, b\rangle\right| \leq \epsilon\ .$$

Thus, also if $N \geq n(\epsilon)$

$$\begin{aligned}
&\left|\frac{1}{N}\sum_{i=1}^{N}\langle U^{k_i}f, g\rangle - \langle f, 1\rangle\langle 1, g\rangle\right| \\
&\leq \left|\frac{1}{N}\sum_{i=1}^{N}\langle U^{k_i}f, g\rangle - \frac{1}{N}\sum_{i=1}^{N}\langle U^{k_i}a, g\rangle\right| \\
&\quad + \left|\frac{1}{N}\sum_{i=1}^{N}\langle U^{k_i}a, g\rangle - \frac{1}{N}\sum_{i=1}^{N}\langle U^{k_i}a, b\rangle\right| \\
&\quad + \left|\sum_{i=1}^{N}\langle U^{k_i}a, b\rangle - \langle a, 1\rangle\langle 1, b\rangle\right| \\
&\quad + \left|\langle a, 1\rangle\langle 1, b\rangle - \langle f, 1\rangle\langle 1, b\rangle\right| \\
&\quad + \left|\langle f, 1\rangle\langle 1, b\rangle - \langle f, 1\rangle\langle 1, b\rangle\right|
\end{aligned}$$

which is

$$\leq \frac{1}{N}\sum_{i=1}^{N}\left|\langle U^{k_i}(f-a),g\rangle\right|$$
$$+\frac{1}{N}\sum_{i=1}^{N}\left|\langle U^{k_i}a,g-b\rangle\right|$$
$$+\ \epsilon\ +\ \left|\langle f-a,1\rangle\langle 1,b\rangle\right|\ +\ \left|\langle f,1\rangle\langle 1,g-b\rangle\right|.$$

Using the fact that $||f||_2 = \langle f,f\rangle^{\frac{1}{2}}$ and Cauchy's inequality this is

$$\leq \epsilon||g||_2 + \epsilon(||f||_2+\epsilon) + \epsilon + \epsilon(||g||_2+\epsilon) + \epsilon||f||_2,$$

thereby completing the proof that Theorem 2.3 (ii) implies Theorem 2.3 (i), as required. ■

Proof that Theorem 2.3 (ii) implies Theorem 2.3 (iv). If $T^{-1}A = A$ then (ii) implies that $\mu(A) = \mu^2(A)$, and so $\mu(A)$ is either zero or one and T is ergodic. In light of Lemma 2.2 this proves Theorem 2.3 (iv). ■

Proof that Theorem 2.3 (iv) implies Theorem 2.3 (v). Let $(N_t)_{t=1}^{\infty}$ denote a strictly increasing sequence of integers such that

$$\left|\left|\frac{1}{N_t}\sum_{i=1}^{N_t} f(T^{k_i}x) - \int_X f(x)d\mu\right|\right|_2 \leq \frac{1}{t}.$$

The existence of such a sequence is ensured by the assumption of Theorem 2.3 (iv). As a consequence we have

$$\sum_{t=1}^{\infty}\left|\left|\frac{1}{N_t}\sum_{i=1}^{N_t} f(T^{k_i}x) - \int_X f(x)d\mu\right|\right|_2^2 < \infty.$$

Rearranging, a step justified by Fatou's lemma, we have

$$\int_X(\sum_{t=1}^{\infty}\left|\frac{1}{N_t}\sum_{i=1}^{N_t} f(T^{k_i}x) - \int_X f(x)d\mu\right|_2^2)d\mu < \infty.$$

We may therefore conclude, using the fact that for a sequence $(a_t)_{t=1}^{\infty}$ of non-negative functions $\sum_{t=1}^{\infty} a_t < \infty$ implies that $a_t = o(1)$, that we have

$$\lim_{t\to\infty} \frac{1}{N_t} \sum_{i=1}^{N_t} f(T^{k_i}x) = \int_X f(x)d\mu(x),$$

almost everywhere with respect to μ. This means that $\ell_{T,f}(x) = \int_X f(x)d\mu$ μ almost everywhere as required. ■

Proof that Theorem 2.3 (v) implies Theorem 2.3 (i). Note that by (v) for any two sets A and B belonging to β we have

$$\lim_{N\to\infty} \frac{1}{N} \sum_{i=1}^{N} \chi_A(T^{k_i}x)\chi_B = \mu(A)\chi_B,$$

almost everywhere with respect to μ. This means that as a consequence of the dominated convergence theorem (i) follows. ■

Proof that Theorem 2.3 (iii) and Theorem 2.3 (iv) are equivalent. Again without loss of generality we assume that $\int_X f(x)d\mu = 0$. The proof of the equivalence of (iii) and (iv) now follows from the following identity.

$$\left|\left|\frac{1}{N}\sum_{i=1}^{N} f(T^{k_i}x)\right|\right|_2^2 = \frac{1}{N^2} \sum_{0\le k_i,k_j\le N} \langle U^{k_i-k_j}f, f\rangle$$

which via the properties of the spectral measure is

$$= \frac{1}{N^2} \sum_{0\le k_i,k_j\le N} \int_{\mathbb{T}} z^{k_i-k_j} d\omega_f(z)$$

$$= \int_{\mathbb{T}} \left|\frac{1}{N}\sum_{i=1}^{N} z^{k_i}\right|^2 d\omega_f(z).$$

This proves that (iii) and (iv) are equivalent if $p = 2$. The case $p \le 2$ follows as before because L^2 is dense in L^p. ■

Proof that Theorem 2.3 (v) implies Theorem 2.3 (iv).

Suppose first that g is bounded and measurable; then in particular g is in L^p and so assuming (v) we have

$$\lim_{N\to\infty} \frac{1}{N} \sum_{i=1}^{N} g(T^{k_i}x) = \int_X g d\mu,$$

almost everwhere with respect to μ. This means that by the bounded convergence theorem we have

$$\lim_{N\to\infty}\left|\left|\frac{1}{N}\sum_{i=1}^{N} g(T^{k_i}x) - \int_X g d\mu\right|\right|_p = 0.$$

Now if we are given $\epsilon > 0$ then there exists a possibly large natural number $n = n(\epsilon, g)$ such that if $N > n$ and k is a positive integer then

$$\left|\left|\frac{1}{N}\sum_{i=1}^{N} g(T^{k_i}x) - \frac{1}{N+k}\sum_{i=1}^{N+k} g(T^{k_i}x)\right|\right|_p < \epsilon.$$

Let us now consider general functions f in $L^p(X, \beta, \mu)$ and set

$$S_N(f) = \frac{1}{N}\sum_{i=1}^{N} f(T^{k_i}x).$$

We want to show that $(S_N(f))_{N=1}^{\infty}$ is a Cauchy sequence in $L^p(X, \beta, \mu)$. First notice that $||S_N(f)||_p \leq ||f||_p$. Suppose we are given $\epsilon > 0$ and that g is in $L^\infty(X, \beta, \mu)$ and that $||f - g||_p < \epsilon$. Then

$$\begin{aligned}||S_N(f) - S_{N+k}(f)||_p \leq\ & ||S_N(f) - S_N(g)||_p \\ & + ||S_N(g) - S_{N+k}(g)||_p \\ & + ||S_{N+k}(g) - S_{N+k}(f)||_p ,\end{aligned}$$

which is less than ϵ, if $N > n(\frac{\epsilon}{2}, g)$. Thus $(S_N(f))_{N=1}^{\infty}$ is a Cauchy sequence, implying that

$$\lim_{N\to\infty}\left|\left|\frac{1}{N}\sum_{i=1}^{N} f(T^{k_i}x) - f^*(x)\right|\right|_p = 0,$$

for some function f^*. But $g^*(x) = \int_X g(x)d\mu$ for g in $L^\infty(X, \beta, \mu)$ and such functions are dense in $L^p(X, \beta, \mu)$ and so $f^*(x) = \int_X f(x)d\mu$ for all functions f in $L^p(X, \beta, \mu)$ as required. Hence Theorem 2.2 (iv) and hence the whole of Theorem 2.2 is proved as required. ■

3. Some consequences for measures on compact metrisable spaces

Throughout this section X will denote a compact metrisable space and $\beta = \beta(X)$ will denote its Borel σ-algebra. For a measurable transformation T of the space X we denote by $M(X,T)$ the set of T invariant probability measures on the measurable space (X,β). Also throughout this section we assume $k = (k_n)_{n=1}^{\infty}$ is both Hartman uniformly distributed and L^p good universal. Suppose also that T is continuous. We denote by $C(X)$ the space of continuous complex-valued functions defined on X. We then have the following theorem.

Theorem 3.1. *Let μ be an element of $M(X,T)$. Then T is ergodic if and only if for each f in $C(X)$ and g in $L^1(X,\beta,\mu)$ we have*

$$\lim_{N\to\infty}\frac{1}{N}\sum_{n=1}^{N}\int_X f(T^{k_n}x)g(x)d\mu(x) = \int_X f d\mu \int_X g d\mu. \tag{3.1}$$

Proof. Suppose first that (3.1) holds for each f in $C(X)$ and each g in $L^1(X,\beta,\mu)$, and let F and G belong to $L^2(X,\beta,\mu)$. In particular G is in $L^1(X,\beta,\mu)$ and so

$$\lim_{N\to\infty}\frac{1}{N}\sum_{n=1}^{N}\int_X f(T^{k_n}x)G(x)d\mu(x) = \int_X f d\mu \int_X G d\mu,$$

for all f in $C(X)$. Now approximate F in $L^2(X,\beta,\mu)$ by continuous functions to get

$$\lim_{N\to\infty}\frac{1}{N}\sum_{n=1}^{N}\int_X F(T^{k_n}x)G(x)d\mu(x) = \int_X F d\mu \int_X G d\mu.$$

Hence (3.1) implies the ergodicity of T. We now show the converse. Suppose the transformation T is ergodic with respect to the space (X,β,μ) for some measure μ in $M(X,T)$. Suppose also that f is an element of $C(X)$ and hence also an element of $L^2(X,\beta,\mu)$. Thus if h is also an element of $L^2(X,\beta,\mu)$ we get

$$\lim_{N\to\infty}\frac{1}{N}\sum_{n=1}^{N}\int_X f(T^{k_n}x)h(x)d\mu(x) = \int_X f d\mu \int_X h d\mu.$$

If g is in $L^1(X,\beta,\mu)$ then by approximating g by elements h of $L^2(X,\beta,\mu)$ we obtain

$$\lim_{N\to\infty}\frac{1}{N}\sum_{n=1}^{N}\int_X f(T^{k_n}x)g(x)d\mu(x) = \int_X fd\mu\int_X gd\mu,$$

as required. ∎

Suppose (X,β,μ,T) is an ergodic transformation. The next application concerns measures m which are absolutely continuous with respect to μ. For a measure m its pullback $T^{-1}m$ under the transformation T is defined by $T^{-1}m(A) = m(T^{-1}A)$ for all A in β where as before $T^{-1}A = \{x : Tx \in A\}$.

Theorem 3.2 : *Suppose the measure μ is an element of $M(X,T)$. Then (X,β,μ,T) is ergodic if and only if whenever m is in $M(X,T)$ and m is absolutely continuous with respect to μ then*

$$\lim_{N\to\infty}\frac{1}{N}\sum_{n=1}^{N}T^{-k_n}m = \mu, \tag{3.2}$$

weakly in $M(X)$.

Proof. First suppose (X,β,μ,T) is ergodic and suppose m in $M(X)$ is absolutely continuous with respect to μ. Let g denote the Radon-Nikodym derivative $\frac{dm}{d\mu}$ which is contained in $L^1(X,\beta,\mu)$. If f is in $C(X)$ then

$$\begin{aligned}\lim_{N\to\infty}\int_X fd(\frac{1}{N}\sum_{n=1}^{N}T^{-k_n}m) &= \lim_{N\to\infty}\frac{1}{N}\sum_{n=1}^{N}\int_X f\circ T^{k_n}dm\\ &= \lim_{N\to\infty}\frac{1}{N}\sum_{n=1}^{N}\int_X f(T^{k_n}x)g(x)d\mu\\ &= \int_X fd\mu\int_X gd\mu\,.\end{aligned}$$

Thus

$$\lim_{N\to\infty}\frac{1}{N}\sum_{n=1}^{N}\int_X f(T^{k_n}x)dm = \int_X fd\mu$$

which is (3.2) as the theorem promises. We now consider the converse. Suppose (3.2) holds. Let g be an element of $L^1(X, \beta, \mu)$ and first suppose that g is positive everywhere. Set m to be the measure defined on X by $m(B) = c\int_X d\mu$ where $c = \frac{1}{\int_X g d\mu}$. Then if f is in $C(X)$ we have

$$= \lim_{N\to\infty} \frac{1}{N}\sum_{n=1}^{N}\int_X f(T^{k_n}x)g(x)d\mu = \int_X f d\mu \int_X g d\mu.$$

Now suppose g is in $L^1(X, \beta, \mu)$ and real valued, write it $g = g_+ - g_-$ where g_+ and g_- are the positive and negative parts of g respectively. The desired conclusion for g follows by applying the deduction of the previous paragraph to g_+ and g_- in turn. The case of complex valued g is treated similarly. ■

The following theorem says that in the context of continuous functions on X there is a universal convergence set.

Theorem 3.3. *If μ is in $M(X,T)$ for the continuous map T of X and (X, β, μ, T) is ergodic, then there exists a set Y in β with $\mu(Y) = 1$ such that for each f in $C(X)$ and each y in Y we have*

$$\lim_{N\to\infty} \frac{1}{N}\sum_{n=1}^{N} f(T^{k_n}y) = \int_X f d\mu.$$

Proof. Choose a countable dense subset $(f_k)_{k=1}^{\infty}$ of $C(X)$. Using Theorem 3.1 there is a set X_k in β such that $\mu(X_k) = 1$ and

$$\lim_{N\to\infty} \frac{1}{N}\sum_{n=1}^{N} f_k(T^{k_n}y) = \int_X f_k d\mu,$$

for all x in X_k. Put $Y = \cap_{k\geq 1} X_k$. We have $\mu(Y) = 1$ and

$$\lim_{N\to\infty} \frac{1}{N}\sum_{n=1}^{N} f_k(T^{k_n}y) = \int_X f_k d\mu,$$

for all y in Y and k in N. The theorem now follows by approximating a given f in $C(X)$ by members of $(f_k)_{k=1}^{\infty}$. ■

The following lemma relates delta measures to ergodicity.

Theorem 3.4. *Suppose T is a continuous map of X and that μ is in $M(X,T)$. Let δ_y denote the delta measure at y; that is, for each set A in β $\delta_y(A) = 1$ if y is an element of A and $\delta_y(A) = 0$ otherwise. Then (X,β,μ,T) is ergodic if and only if we have*

$$\lim_{N\to\infty}\frac{1}{N}\sum_{n=1}^{N}\delta_{T^{k_n}y} = \mu,$$

weakly μ almost everywhere in y.

Proof. If (X,β,μ,T) is ergodic then by Theorem 3.3 we immediately have

$$\lim_{N\to\infty}\frac{1}{N}\sum_{n=1}^{N}\delta_{T^{k_n}y} = \mu,$$

weakly in $M(X)$ for all y in Y, where $\mu(Y) = 1$. Conversely suppose

$$\lim_{N\to\infty}\frac{1}{N}\sum_{n=1}^{N}\delta_{T^{k_n}y} = \mu,$$

weakly for all y in Y with $\mu(Y) = 1$. Then for all f in $C(X)$ we have

$$\lim_{N\to\infty}\frac{1}{N}\sum_{n=1}^{N}f(T^{k_n}y) = \int_X f\,d\mu,$$

for all y in Y. If y is in Y and f is in $C(X)$ and g is in $L^1(X,\beta,\mu)$ then

$$\lim_{N\to\infty}\frac{1}{N}\sum_{n=1}^{N}f(T^{k_n}y)g(y) = g(y)\int_X f\,d\mu,$$

and so, integrating both sides over Y with respect to μ, Theorem 3.1 yields the ergodicity of (X,β,μ,T) as required. ∎

Recall that we say that (X,β,μ,T) is uniquely ergodic if the only element in $M(X,T)$ is μ. Necessarily this means (X,β,μ,T) is an ergodic transformation. We need a lemma.

Lemma 3.5. *The following statements are equivalent :*

(a) the transformation (X, β, μ, T) is uniquely ergodic ;

(b) for each continuous function f defined on X there is a constant C_f independent of x such that

$$\lim_{N\to\infty} \frac{1}{N} \sum_{n=1}^{N} f(T^{k_n} x) = C_f,$$

uniformly on X; and

(c) whenever f is in $C(X)$

$$\lim_{N\to\infty} \frac{1}{N} \sum_{n=1}^{N} f(T^{k_n} x) = \int_X f d\mu,$$

pointwise on X.

Proof. We first consider the proof that (c) implies (a). Let

$$S_N f(x) = \frac{1}{N} \sum_{n=1}^{N} f(T^{k_n} x) . \qquad (N = 1, 2, \ldots)$$

For ν in $M(X,T)$, by the dominated convergence theorem we have

$$\lim_{N\to\infty} \int_X S_N f(x) d\nu(x) = \int_X f d\nu = \int_X f d\mu .$$

This holds for all f in $C(X)$ and hence by the Riesz representation theorem we have $\nu = \mu$, as required.

We now show how (a) implies (b). Suppose (b) does not hold. Then there exists an $\epsilon > 0$, a function g in $C(X)$ and a sequence $(x_{n_j})_{j=1}^{\infty}$ in X such that

$$\left| S_{n_j} g(x_{n_j}) - \int_X g d\mu \right| > \epsilon .$$

Using the unique ergodicity property of (X, β, μ, T) and refining $(x_{n_j})_{j=1}^{\infty}$ if necessary we can find ν in $M(X,T)$ such that

$$\left| \int_X g d\nu - \int_X g d\mu \right| \geq \epsilon ,$$

which is a contradiction.

The proof that (b) implies (a) is obvious. ■

We are now ready to prove the final theorem of this section. For a transformation (X, β, μ, T) we say a point x in X is k generic if

$$\lim_{N\to\infty} \frac{1}{N} \sum_{n=1}^{N} f(T^{k_n} x) = \int_X f d\mu ,$$

holds for all functions f in $C(X)$. Recall that we say a continuous transformation T of X is minimal if the only closed subsets it leaves invariant are X and the empty set.

Theorem 3.6. *Suppose T is minimal on X and assume for each x in X there exists a measure μ_x such that x is k generic for (X, β, μ, T). Then (X, T) is uniquely ergodic.*

Proof. By assumption, for every x in X

$$\lim_{N\to\infty} \frac{1}{N} \sum_{n=1}^{N} f(T^{k_n} x) = \int_X f d\mu_x = C(f, x).$$

We have to show that $C(f, x)$ is independent of x. Fix f in $C(X)$. Then the function $C(f, x)$ is the pointwise limit of continuous functions and T invariant because of the Hartman distribution of k. Because of the assumed minimality, for x_0 in X the sequence $(T^{k_n} x_0)_{n=1}^{\infty}$ is dense in X and so if x_0 were a point of discontinuity, then $C(f, x)$ would be nowhere continuous contradicting the Baire category theorem which tells us that the points of discontinuity of $C(f, x)$ are of first category. This means that $C(f, x)$ is constant as required. ■

References

[AN] N. H. Asmar and R. Nair: *Certain averages on the a-adic numbers*, Proc. Amer. Math. Soc. **114** no. 1 (1992), 21–28.

[Bo1] J. Bourgain: *On the maximal ergodic theorem for certain subsets of the integers*, Israel J. Math. **61** (1988), 39–72.

[Bo2] J. Bourgain: *Pointwise ergodic theorems for arithmetic sets*, Publ. I. H. E. S. **69** (1989), 5–45.

[H] S. Hartman: *Remarks on equidistribution on non-compact groups*, Compositio Math., **16** (1964), 66–71.

[Ka] Y. Katznelson: *An introduction to Harmonic Analysis* , Wiley, (1968).

[Kr] U. Krengel: *Ergodic Theorems*, de Gruyter Studies in Mathematics **6** (1985).

[KN] L. Kuipers and H. Neiderreiter: *Uniform Distribution of Sequences*, Wiley, (1974).

[Na1] R. Nair: *On polynomials in primes and J. Bourgains circle method approach to ergodic theorems*, Ergododic Theory and Dynamical Systems, **11** (1991), 485–499.

[Na2] R. Nair: *On polynomials in primes and J. Bourgains circle method approach to ergodic theorems II*, Studia Math. **105** (3) (1993), 207–233.

[Na3] R. Nair: *On uniformly distributed sequences of integers and recurrence*, (preprint).

[Na4] R. Nair: *On uniformly distributed sequences of integers and recurrence II*, (preprint).

[N] I. Niven: *Uniform distribution of sequences of integers*, Trans. Amer. Math. Soc. **98** (1961), 52–61.

[W] P. Walters: *An Introduction to Ergodic Theory*, Springer-Verlag, Graduate Texts in Mathematics **79** (1981).

Department of Pure Mathematics, University of Liverpool, P.O. Box 147, Liverpool L69 3BX, UK

Hypergroups and Signed Hypergroups

Kenneth A. Ross

1. Introduction to Hypergroups

Hypergroups, as I understand them, have been around since the early 1970's when Charles Dunkl, Robert Jewett and René Spector independently created locally compact hypergroups with the purpose of doing standard harmonic analysis. As one would expect, there were technical differences in their definitions. The standard, in the non-Soviet world, became Jewett's 101-page paper [J] because he worked out a good deal of the basic theory that people would want. Bloom and Heyer's book [BH] is a useful report on some of the mathematics that has been done on the basis of Jewett's axioms.

In August of 1993 the first conference on hypergroups was held in Seattle. This first conference was an international conference and was well attended by people from all over the world. See [CGS]. As the preface to the proceedings states, "This led to fireworks. Hypergroups occur so often and in so many different and important contexts, that mathematicians all over the world have been discovering the same mathematical structure hidden in very different applications, and publishing theorems about these structures, in many cases without even knowing that they were talking about hypergroups."

In particular, structures very like the Dunkl-Jewett-Spector creations of the 1970's had been studied in the early 1950's by Berezansky and colleagues. The axioms, terminology and language were all different, and the connection was not realized by most workers until the Seattle conference in 1993. These connections are explained in the nice article [BKl], where it is noted that the ideas of hypergroups appear in works of Delsarte and Levitan published in 1938 and 1940. See also the very recent article [BK2]. A fundamental part of their axioms are their "structure measures" $c(A, B, r)$ which for locally compact groups with Haar measure m reduce to $m((A - r) \cap B)$. I think the axioms involving $c(A, B, r)$, which have no counterpart in the axioms of a locally compact group, are unnatural compared to the axioms of Jewett and others. In any case, in the proceedings [CGS] of the conference there was an effort to standardize notation. In many of the papers, the hypergroups studied by Jewett [and used in the book of Bloom and

Heyer] are referred to as DJS-hypergroups. These are the hypergroups that I will be talking about.

Incidentally, there was a follow-up conference at Oberwolfach in 1994 that was organized by Herbert Heyer. The proceedings [H] contains many interesting articles, including several on hypergroups.

Here are the axioms for a DJS-hypergroup. As I learned from [K], they have been neatly rephrased by Lasser [L], so I will give Lasser's version. We begin with a locally compact Hausdorff space K. $M(K)$ will denote the space of all finite complex regular measures on K, and $M^1(K)$ will denote the probability measures in $M(K)$. Point masses will be denoted by δ_x. A hypergroup is determined by K and the following data:

(H^*) A continuous mapping $(x, y) \to \delta_x * \delta_y$ from $K \times K$ into $M^1(K)$, where $M^1(K)$ has the weak topology with respect to the space $C_c(K)$ of continuous complex-valued functions with compact support. [convolution]

($H^\vee$) An involutive homeomorphism $x \to x^\vee$ from K to K. [an involution]

(He) A fixed element e in K. [an identity element]

After identification of x with δ_x, the mapping in (H^*) has a unique extension to a continuous bilinear mapping $(\mu, \nu) \to \mu * \nu$ from $M(K) \times M(K)$ into $M(K)$. And the involution on K gives an involution $\mu \to \mu^*$ on $M(K)$, where $\mu^*(E)$ is the conjugate of $\mu(E^\vee)$ for each Borel set E in K.

Now a DJS-hypergroup is the quadruple $(K, *, \vee, e)$ satisfying

(H1) $\delta_x * (\delta_y * \delta_z) = (\delta_x * \delta_y) * \delta_z$ for all x, y, z in K.

(H2) $(\delta_x * \delta_y)^\vee = \delta_{y^\vee} * \delta_{x^\vee}$ for all x, y in K.

(H3) $\delta_x * \delta_e = \delta_e * \delta_x = \delta_x$ for all x in K.

(H4) e is in the support, $\text{supp}(\delta_x * \delta_{y^\vee})$, if and only if $x = y$.

(H5) $\text{supp}(\delta_x * \delta_y)$ is compact for all x, y in K.

(H6) The mapping $(x, y) \to \text{supp}(\delta_x * \delta_y)$ of $K \times K$ into the space of nonvoid compact subsets of K is continuous, where the latter space is given the "Michael" topology in [J], §2.5.

A locally compact group G is a hypergroup satisfying these axioms. In this case, $\delta_x * \delta_y = \delta_{xy}$ for all x, y in G and $x^\vee$ is the inverse of x. Axioms (HI)–(H4) are clear. Note that axiom (H4) says that $xy^{-1} = e$ if and only if $x = y$. Axiom (H5) is very clear since each $\text{supp}(\delta_x * \delta_y)$ consists of the single element xy. The technical axiom (H6) can also be verified in this setting.

I will only give one class of examples at the end of this section, but many examples appear throughout this publication.

A very key concept in locally compact groups and hypergroups is that of an invariant measure. For the group case, Alfred Haar proved in 1933 that a (second countable) locally compact group has a left-invariant measure m, which we now call a left Haar measure. Later authors proved that the left Haar measure is unique up to a positive constant. Left invariant means that $m(xE) = m(E)$ for all x in G and all Borel sets E in G. This is equivalent to the requirement that $\delta_x * m = m$ for all x in G, where the convolution here is extended in the natural way so as to apply even if m is an infinite measure. Similarly, a right Haar measure m' is one that satisfies $m' * \delta_x = m'$ for all x in G. Of course, locally compact groups also have right-invariant Haar measures. We will return to this topic.

As in the group case, a measure m on a DJS-hypergroup K is called a left Haar measure if $\delta_x * m = m$ for all x in K, with a similar definition for right Haar measure. Does every K have Haar measures? Each of the pioneers Dunkl, Jewett and Spector proved that every compact hypergroup K has a left Haar measure. I believe that each of them proved the same results for discrete hypergroups. In any case, the result is easy. The formula is $m(\{x\}) = [\delta_{x^\vee} * \delta_x](\{e\})^{-1}$ for all x in K.

The case for commutative hypergroups was substantially harder, but Spector [S] proved that every commutative hypergroup has a Haar measure. Incidentally, Spector uses weaker axioms than the DJS-axioms and pays a price. He doesn't assume axiom (H4). His proofs simplify somewhat if we use the DJS-axioms. Remarkably, the general question of whether every DJS-hypergroup has a Haar measure is still open. This is certainly the biggest open question in the subject. Note that, just as in the group case, when the Haar measures exist they are unique up to a constant. This was shown by Jewett [J].

As in the group case, we say that a hypergroup is unimodular if the left and right Haar measures coincide. Groups that are either commutative, compact, or discrete are unimodular. Obviously commutative hypergroups are unimodular, and it is not hard to show that compact hypergroups are also unimodular. Surprisingly, on page 39 of Bloom and Heyer's book it is noted that "in contrast to the group case it is unknown whether all discrete hypergroups are necessarily unimodular." However, the recent paper [KW] contains a construction for a class of non-unimodular discrete hypergroups! They arise as double-coset hypergroups induced by the transitive action of non-unimodular groups of permutations on an infinite set.

The most work has been done on commutative hypergroups for the simple reason that they are easier to deal with. As in the case of groups, Fourier and Fourier–Stieltjes transforms play a big role. These are functions that, in the group case, are defined on the character group. In the case of a hypergroup K, they are defined on the space $K^\wedge$ of all characters, which might or might not be a hypergroup in its own right. There are even three-element hypergroups K for which $K^\wedge$ is not a hypergroup. This is an interesting issue and, for finite hypergroups, I will touch on it again in section 3.

A hypergroup character χ on a hypergroup K is a bounded complex-valued continuous function that is not identically zero and satisfies

(i) $\chi(x^\vee)$ is the conjugate of $\chi(x)$ for all x in K,

(ii) $\chi(x * y) = \chi(x)\chi(y)$ for all x, y in K.

Each of these requirements deserves comment. With the other requirements, (i) holds automatically for locally compact groups because $xx^{-1} = e$ for all x. This remark is also true for compact hypergroups, but in general there are functions χ that satisfy (ii) and not (i). Property (ii) needs clarification because $x * y$ isn't defined. This very suggestive notation is due to Jewett. In general,

$$f(x * y) = \int_K f d(\delta_x * \delta_y).$$

In the group case, this is $f(xy)$ just like it should be.

Many familiar results carry over to general and to commutative hypergroups. I will only mention the Levitan–Parseval identity, which I will state in the easy case that K is finite and commutative. Then $K^\wedge$ has a Plancherel measure ν, and

Parseval's identity. $$\sum_{x \in K} |f(x)|^2 m(x) = \sum_{\chi \in K^\wedge} |f^\wedge(\chi)|^2 \nu(\chi).$$

If m is normalized so that $m(K) = 1$, then ν is normalized so that $\nu(1) = 1$.

Let me end with a family of very simple hypergroups that will play a role in the next section. Here are **all** of the two-element hypergroups. For $0 < \theta \leq 1$, the set K will be $\{0, 1\}$, 0 will serve as the identity and the involution will be the identity map. Since 0 is the identity, the three products $\delta_0 * \delta_0 = \delta_0$, $\delta_0 * \delta_1 = \delta_1 * \delta_0 = \delta_1$ are automatic. The only interesting product is

$$\delta_1 * \delta_1 = \theta\delta_0 + (1-\theta)\delta_1.$$

Since the coefficients have to be nonnegative and add to 1, we must have $0 \le \theta \le 1$. Since 0 has to be in the support of $\delta_1 * \delta_1$, we have to have $\theta > 0$. It is easy to check that we get a hypergroup for each θ. Note that for $\theta = 1$, we get the familiar two-element group $Z(2)$.

$Z_\theta(2)$ has two characters, the identically 1 function and χ where $\chi(0) = 1$ and $\chi(1) = -\theta$. The normalized Haar measure m_θ is given by $m_\theta(0) = \theta/(1+\theta)$ and $m_\theta(1) = 1/(1+\theta)$. Again, note that if $\theta = 1$, we get the correct characters and Haar measure on $Z(2)$. If we think of θ as varying continuously starting at 1, then we can view the family $\{Z_\theta(2) : 0 < \theta \le 1\}$ of hypergroups as a deformation of the group $Z(2)$.

In the next section I will focus on the group $Z(2)^d$ of all d-tuples of 0's and 1's. This can be deformed into a big product of hypergroups, namely $Z_\theta(2)^d$. The hypergroup characters of $Z_\theta(2)^d$ are easy to determine in terms of the characters of each factor $Z_\theta(2)$. And the Haar measure m_θ on $Z_\theta(2)^d$ is just the product of the Haar measures on each factor. Thus

$$m_\theta(x) = \theta^{d-H(x)}/(1+\theta)^d \quad \text{for all} \quad x \quad \text{in} \quad Z_\theta(2)^d.$$

Remember, x is a string of 0's and 1's; here $H(x)$ is the number of these terms that are equal to 1.

2. Markov Chains and Deformations of Hypergroups

I am going to begin by discussing "random walks." The walk needs to be governed by some rules: initially our walks will be in finite groups. Random, as the word suggests, means that at each step the walk will be governed by a probability Q. I will assume that a particle starts at the identity of the group. Q will describe the various first steps and their probabilities. $Q^{(n)}$ will describe the position of the particle after n steps. It turns out that $Q^{(n)}$ is the convolution of Q n-times.

Let U be the uniform probability measure on a finite group G. This is, of course, Haar measure normalized to give the group total mass 1. It has been known since at least the 1940's that $Q^{(n)}$ converges to U unless there are obvious impediments, like the support of Q lies in a proper subgroup of G. In other words, if there were many particles, the particles would be evenly mixed after awhile. In the past fifteen years, under the guidance of the guru Persi Diaconis, there has been renewed interest in studying how fast $Q^{(n)}$ converges to U. First, one needs a measurement of closeness. Here's a commonly accepted measurement, called the total variation distance:

$$\|Q^{(n)}-U\| = \max\{|Q^{(n)}(A)-U(A)| : A \subset G\} = \frac{1}{2}\sum_{y\in G} |Q^{(n)}(y)-U(y)|.$$

Here are some examples.

(a) Repeated shuffling of a deck of cards can be viewed as a random walk on the large non-abelian group of all permutations of the deck. If one shuffles with the same randomness, then there is a unique probability Q that describes the shuffles.. Diaconis and his colleagues have studied the difficult question of determining the rate at which $\|Q^{(n)} - U\|$ converges to 0. This depends, of course, on the choice of Q but there are interesting choices of Q and choices that reflect real shuffles. The papers [BD] and [AD] provide nice introductions to this subject. The book [D] is the basic "textbook" of the subject.

(b) Interesting random walks occur in much simpler abelian groups, like the cyclic group $Z(q)$ on q elements. One natural Q is the "nearest neighbor random walk" where $Q(1) = Q(q-1) = 0.5$. If q is even, then this lives on a proper subgroup. To avoid this sort of problem, we sometimes study the "nearest neighbor or stay at home random walk." Here $Q(1)$, $Q(q-1)$, and $Q(0)$ may all be set equal to one-third.

Before going on, let me mention a powerful, but simple, tool that Diaconis and his co-workers have used. They call it the Upper Bound Lemma. They have a version for non-abelian groups, but here I will only consider finite abelian groups. It is an easy consequence of Parseval's identity after bounding the norm by an ℓ^2 norm.

Upper Bound Lemma. $4\|Q^{(n)} - U\|^2 \leq \sum_{\chi\neq 1} |Q^\wedge(\chi)|^{2n}$.

The easiest non-trivial application is to $Z(3)$ with the nearest neighbor random walk $Q(1) = Q(2) = 0.5$. We worked this out in [RX3] and found that $\|Q^{(n)} - U\|^2 \leq 2^{-2n-1}$. For $n = 10$, this yields $\|Q^{(10)} - U\| \leq 0.000691$. Direct calculation shows that $\|Q^{(10)} - U\|$ is approximately 0.000650.

(c) Now consider the cube $Z(2)^d$ of all d-tuples of 0's and 1's. For each x in $Z(2)^d$, let $H(x)$ be the number of coordinates of x with a 1. This is the Hamming distance from x to the origin. There's a natural graph of the group where one connects two elements if they differ in exactly one coordinate, i.e., if $H(x-y) = 1$. A natural random walk is the "nearest neighbor random walk" where each of the d elements with $H(x) = 1$ has probability $1/d$. The only problem is that there can be parity problems. To avoid these we again use the "nearest neighbor or stay at home" random walk. I.e., $Q(0) = 1/(d+1)$ and $Q(x) = 1/(d+1)$ for all x with $H(x) = 1$. The Upper Bound Lemma can be used to estimate $\|Q^{(n)} - U\|$. Some combinatorial estimates are needed. This quantity goes to 0 exponentially with n. For example, if $d = 2$, we find that $4\|Q^{(n)} - U\|^2 \leq (1/3)^{2n-1}$.

We can view every random walk as a Markov chain. In fact, the transition matrix M corresponding to Q is given by $M(x,y) = \delta_x * Q(y)$ for all x, y in G. Here δ_x denotes the point mass at x and $*$ denotes convolution. Note that $M(e,y) = Q(y)$, where e is the identity of the group. It follows easily by induction that the matrix power satisfies $M^n(x,y) = [\delta_x * Q^{(n)}](y)$ for x, y in G. If we want to study Markov chains, there are many techniques for studying them. However, if it turns out that the Markov chain is actually a random walk, then we also have the tools of elementary harmonic analysis at our disposal, as illustrated by the use of the Upper Bound Lemma.

Diaconis and Hanlon [DH] studied some special Markov chains called Metropolis Markov chains. As I will explain, the ones that they studied are based on familiar random walks on finite groups, but are not themselves random walks. Metropolis Markov chains are so-named because of an old paper with four or five authors, among them Nicholas Metropolis and Edward Teller.

Metropolis Markov chains make sense on any finite set X. Let π be a probability on X with $\pi(x) > 0$ for all x in X. The Metropolis algorithm is a classical Markov chain simulation method for sampling from π, which is effective when the ratios $\pi(x)/\pi(y)$ are available. We begin with a "base chain" $B(x,y)$ that is partly symmetric, i.e., $B(x,y) = 0$ if and only if $B(y,x) = 0$. For us, B will be generated by a random walk. We will use the ratios

$$r_{y,x} = [\pi(y)B(y,x)]/[\pi(x)B(x,y)],$$

where we decree that this is 0 if $B(x,y) = B(y,x) = 0$. Here is the Metropolis Markov chain:

$$\begin{aligned} M(x,y) &= B(x,y) && \text{if } y \neq x \text{ and } r_{y,x} \geq 1; \\ &= B(x,y) r_{y,x} && \text{if } r_{y,x} < 1; \\ &= B(x,x) + E(x) && \text{if } y = x, \end{aligned}$$

where $E(x) = \sum B(x,z)(1 - r_{z,x})$, summed over all $z \neq x$ with $r_{z,x} < 1$.

That is, $E(x)$ is the value needed so that the sum of all $M(x, y)$, over y in X, is 1.

The stationary distribution for this Markov chain will be the original probability π. Diaconis and Hanlon were interested in the convergence rates of these Markov chains, but note that now this may well depend on the starting point. So the variation distance is defined by

$$\|M^n(x, \_) - \pi\| = (1/2) \sum |M^n(x, y) - \pi(y)|.$$

Consider again the nearest neighbor walk on $Z(2)^d : Q(x) = 1/d$ if and only if $H(x) = 1$. Note that the corresponding Markov chain B is given by

$$B(x, y) = \delta_x * Q(y) = Q(y - x) = 1/d \text{ if and only if } H(x - y) = 1.$$

All we need now, to specify a Metropolis Markov chain, is the stationary distribution. Diaconis and Hanlon consider the measures π_θ defined on $Z(2)^d$ by

$$\pi_\theta(x) = \theta^{H(x)}/(1 + \theta)^d, \quad \text{where} \quad 0 < \theta \leq 1.$$

When $\theta = 1$, this is just the uniform probability measure on the group $Z(2)^d$. Diaconis and Hanlon find the eigenvectors and eigenvalues of the Metropolis Markov chain M in terms of Krawtchouk polynomials and then establish the rate of convergence of $\|M^n(\mathbf{0}, \_) - \pi_\theta\|$. I won't give you the complicated results, but this will be small if n is roughly $d \log(d) + c$, where c does not depend on d.

In [RX1], Daming Xu and I analyzed the Diaconis–Hanlon result from a little different point of view. First, recall that the hypergroups $Z_\theta(2)^d$ have Haar measure given by $m_\theta(x) = \theta^{d - H(x)}/(1 + \theta)^d$. This is very like the π_θ studied by Diaconis–Hanlon. Since $H(x)$ counts the number of 1's in x, and $d - H(x)$ counts the number of 0's in x, a simple change of variable, $x \to \mathbf{1} - x$, in the Diaconis–Hanlon Markov chain changes their stationary distribution to the Haar measure. With this trivial change, we see that they estimated the rate of convergence of $\|M^n(\mathbf{1}, \_) - m_\theta\|$.

Since random walks on a finite group converge to the Haar measure of the group, unless there are obvious impediments, this suggests that perhaps M is, in fact, a random walk but **on the hypergroup** $Z_\theta(2)^d$. This is the case and the probability measure that generates this random walk is exactly the nearest neighbor random walk Q. The Metropolis Markov chains studied by Diaconis and Hanlon, which can be viewed as deformations of the nearest neighbor random walk, are precisely the nearest neighbor random walks on the hypergroup deformations of the group $Z(2)^d$.

The Upper Bound Lemma mentioned before carries over to hypergroups with no difficulty. It now reads

Upper Bound Lemma. $4\|Q^{(n)} - m\|^2 \leq \sum_{\chi\neq 1} |Q^\wedge(\chi)|^{2n}\nu(\chi)$.

Here ν is the Plancherel measure on $K^\wedge$. Since ν is counting measure in the group case, this result is an exact generalization of the original Upper Bound Lemma. For the nearest neighbor random walk Q on the hypergroup $Z_\theta(2)^d$, the summands in the Upper Bound Lemma are easily calculated. Some careful algebraic estimations then lead to exactly the same bounds that Diaconis and Hanlon obtained. We get one benefit, though, because it is now clear that the bounds work no matter what point our random walk starts at. That is, the bounds that Diaconis and Hanlon obtained for $\|M^n(\mathbf{0},\_)-\pi_\theta\|$ hold for $\|M^n(x,\_)-\pi_\theta\| = \|M^n(\mathbf{1}-x,\_) - m_\theta\|$.

Using the same methods, Daming Xu and I also obtained similar estimates for the Metropolis Markov chain associated with the nearest neighbor random walk on $Z(3)^d$. In [RX2], we work hard to obtain similar results involving the groups S_n of all permutations of an n-element set. Again the basic random walk is the nearest neighbor random walk starting at the identity permutation. The nearest neighbors are transpositions. Diaconis and Hanlon studied the Metropolis Markov chain in this setting. They "lumped" the chain to the space K_n of conjugacy classes. The Markov chains so obtained are still not random walks, but they are if we then deform K_n into an object that isn't even a hypergroup. The objects are called signed hypergroups. They will be the topic in section 3. I won't return to the messy calculations necessary to establish rates of convergence, but I want to emphasize that our interest in signed hypergroups began here.

[RX3] contains a nice expository account of all of this, and much more. In particular, we work out the details for some Metropolis Markov chains on S_3^d and on S_4^d. Note that this article was published in the J. Math. Sciences in honor of my friend U. N. Singh.

3. Signed Hypergroups

As we all know, the theory of locally compact hypergroups is well established. The theory is quite rich and many generalizations from locally compact groups have been obtained. But even more general structures have been studied and found to be useful. Indeed, much of the work of Berezansky and other former Soviet mathematicians were in a more general context. However, in this section I am going to focus on what are called signed hypergroups. My focus will be on axiomatics. The process is still ongoing in the sense that I still do not know the "right" axioms for a general locally compact (Hausdorff) signed hypergroup.

Recall that, in section 2, I indicated that my colleague Daming Xu and I were led to signed hypergroups because we were looking for ways to look at certain Markov chains as random walks. We were fortunate to learn about a very simple elegant system of axioms, due to Norman Wildberger [W], for what are now called finite signed hypergroups. This story is told in my Seattle conference survey [Ro], so I will only give enough here to make this report coherent. Let's look again at the key axioms for a DJS-hypergroup. We are given a locally compact Hausdorff space K with an involutive homeomorphism $X \to x^\vee$ and a special element e. In addition

(H*) A continuous mapping $(x, y) \to \delta_x * \delta_y$ from $K \times K$ into $M^1(K)$, where $M^1(K)$ has the weak topology with respect to the space $C_c(K)$ of continuous complex-valued functions with compact support. [convolution]

(H1) $\delta_x * (\delta_y * \delta_z) = (\delta_x * \delta_y) * \delta_z$ for all x, y, z in K.

(H2) $(\delta_x * \delta_y)^\vee = \delta_{y^\vee} * \delta_{x^\vee}$ for all x, y in K.

(H3) $\delta_x * \delta_e = \delta_e * \delta_x = \delta_x$ for all x in K.

(H4) e is in the support $\text{supp}(\delta_x * \delta_{y^\vee})$ if and only if $x = y$.

(H5) $\text{supp}(\delta_x * \delta_y)$ is compact for all x, y in K.

(H6) The mapping $(x, y) \to \text{supp}(\delta_x * \delta_y)$ of $K \times K$ into the space of nonvoid compact subsets of K is continuous, where the latter space is given the "Michael" topology in [J], §2.5.

For a finite DJS-hypergroup, axioms (H5) and (H6) are not needed since they automatically hold. This leaves us with (H*) and (H1)–(H4), and the continuity condition in (H*) is no longer an issue. The only change needed for finite signed hypergroups is to weaken (H*) to

(SH*) For each (x, y) in $K \times K$, $\delta_x * \delta_y$ is a real (or signed) measure satisfying $\delta_x * \delta_y(K) = 1$,

and to add

(Pe) $\delta_{x^\vee} * \delta_x(e) > 0$ for all x in K.

We need axiom (Pe) in order to show the existence of Haar measure. It's given by the same formula I mentioned for discrete hypergroups:

$m(\{x\}) = [\delta_{x^\vee} * \delta_x](\{e\})^{-1}$ for all x in K.

I should mention right now that Margit Rösler weakens (SH) further and only requires that $\delta_x * \delta_y$ be a real measure. However, she shows in her setting that she gets an invariant measure if and only if $\delta_x * \delta_y(K) = 1$. She weakens the hypothesis because in [R1] she studies a Laguerre convolution system and in [R2] she studies Bessel-type signed hypergroups that do not have this property. See [R3] and [R4] for more recent papers by Rösler on signed hypergroups.

A merit of Wildberger's axioms is the following.

Duality theorem. *If K is a finite commutative signed hypergroup, then $K^\wedge$ is also a commutative signed hypergroup under pointwise operations and conjugation, and K is the character hypergroup of $K^\wedge$ in a natural way. In particular, if K is a finite commutative hypergroup, then even though $K^\wedge$ need not be a hypergroup, it is a signed hypergroup and its dual is K.*

Moreover, one can do "harmonic analysis" in the setting of these signed hypergroups. I would like to see the correct axioms for general locally compact signed hypergroups. For one thing, they should be axioms that, when each $\delta_x * \delta_y$ is a probability measure, reduce to the axioms of a DJS-hypergroup. I will discuss where we seem to stand.

Incidentally, every two-element signed hypergroup has the form $Z_\theta(2) = \{0, 1\}$ where $\delta_1 * \delta_1 = \theta\delta_0 + (1-\theta)\delta_1$, where we now allow θ to be any positive number.

I believe that I know the correct axioms for a discrete signed hypergroup. I believe that we need to reinstate (H5):

(H5) $\operatorname{supp}(\delta_x * \delta_y)$ is compact [i.e., finite] for all x, y in K,

and add

$$\sup\{\|\delta_x * \delta_y\| : x, y \text{ in } K\} \quad \text{is finite.}$$

This is needed if one is going to get convolution inequalities that make L^1 into a Banach algebra, L^p into an L^1-Banach module, etc.; see §4 of [Ro].

Beyond the discrete case, the situation is much less clear. In [Ro] I proposed some axioms for compact signed hypergroups. At the same time, Margit Rösler [R1] proposed a similar set of axioms for σ-compact signed hypergroups. My student Adam Parr [P] is trying to find a good common set of axioms for arbitrary (locally compact) signed hypergroups. The goal is to find axioms so that

(a) for discrete signed hypergroups, they agree with the axioms that

I've given;

(b) when each $\delta_x * \delta_y$ is a nonnegative measure, hence a probability measure, then the axioms agree with Jewett's axioms for DJS-hypergroups;

(c) for σ-compact signed hypergroups, they are as compatible with Margit Rösler's axioms as possible when one adds axioms (SH*) and (H5) to her set of axioms.

I will set down Parr's tentative set of axioms. I will then compare them with Jewett's axioms and with Rösler's axioms. I will also discuss some of the issues involved. Here are Parr's axioms: We are given a locally compact Hausdorff space K with an involutive homeomorphism $x \to x^\vee$, a special element e, and an associative bilinear mapping $(\mu, \nu) \to \mu * \nu$ from $M(K) \times M(K)$ into $M(K)$. In addition

(SH*) A continuous mapping $(x, y) \to \delta_x * \delta_y$ from $K \times K$ to $M(K)$, where $M(K)$ has the weak topology τ_b with respect to the space $C_b(K)$ of all bounded continuous complex-valued functions on K.

(SH1) for each (x, y) in $K \times K$, $\delta_x * \delta_y$ is a real (or signed) measure satisfying $\delta_x * \delta_y(K) = 1$.

(SH2) $(\delta_x * \delta_y)^\vee = \delta_{y^\vee} * \delta_{x^\vee}$ for all x, y in K.

(SH3) $\delta_x * \delta_e = \delta_e * \delta_x = \delta_x$ for all x in K.

(SH4) z is in $\operatorname{supp}(\delta_x * \delta_y)$ if and only if y is in $\operatorname{supp}(\delta_{x^\vee} * \delta_z)$ for all x, y, z in K.

(SH5) $\operatorname{supp}(\delta_x * \delta_y)$ is compact for all x, y in K.

(SH6) The mapping $(x, y) \to \operatorname{supp}(\delta_x * \delta_y)$ of $K \times K$ into the space of nonvoid compact subsets of K is continuous, where the latter space is given the "Michael" topology in [J], §2.5.

(SH7) $\sup\{\|\delta_x * \delta_y\| : x, y \text{ in } K\}$ is finite.

(SH8) There exists a left invariant measure m on K with $\operatorname{supp}(m) = K$ and satisfying

$$(*) \quad \int_K f(x * y) g(y) dm(y) = \int_K f(y) g(x^\vee * y) dm(y) \text{ for all } x \text{ in } K \text{ and } f, g \text{ in } C_c(K).$$

Remarks. (SH*) Jewett's version assumes that the mapping $(x, y) \to \delta_x * \delta_y$ from $K \times K$ to $M^+(K)$ is continuous, where $M^+(K)$ has the weak topology with respect to the space $C_c(K) \cup \{1\}$; he calls this the "cone topology." This topology agrees with τ_b on $M^+(K)$. This also agrees with the weak topology on $M^1(K)$ with respect to $C_c(K)$. Rösler assumes that K is σ-compact and assumes that the mapping $(x, y) \to \delta_x * \delta_y$ from $K \times K$ to $M(K)$ is continuous, where $M(K)$ has the weak topology with respect to the space $C_0(K)$ of the bounded continuous complex-valued functions on K that vanish at infinity. Whatever the topology is on $M(K)$, the goal is to have the convolution mapping $(\mu, \nu) \to \mu * \nu$ continuous or separately continuous on some domain. Jewett gets it continuous in the cone topology from $M^+(K) \times M^+(K)$ to $M^+(K)$. Rösler gets it separately continuous from $M(K) \times M(K)$ to $M(K)$ by adding another axiom: given f in $C_c(K)$ and x in K, each map $y \to f(x * y)$ and $y \to f(y * x)$ is in $C_c(K)$.

Parr gets the mapping $(\mu, \nu) \to \mu * \nu$ continuous from $M(K) \times M(K)$ to $M(K)$. He used the strict topology on $C_b(K)$ as a tool, and his work was motivated by help from Professor Ajit Iqbal Singh.

(SH1) As mentioned before, Rösler does not require $\delta_x * \delta_y(K) = 1$ because she has important examples without this property. Of course, there isn't an invariant measure whenever this property fails.

(SH2) With the current set of axioms, this one is redundant. Rösler [R3] points out that (SH2) is a consequence of (*) in axiom (SH8).

(SH4) Note that, in the group case, this axiom merely states that $z = xy$ if and only if $y = x^{-1}z$. This axiom immediately implies axiom (H4), and the two axioms are equivalent for discrete signed hypergroups. Rösler does not assume either of these axioms. With her weaker axioms, in [R3] she is able to prove that if x is in X and e does not belong to any $\text{supp}(\delta_x * \delta_y)$ for $y \neq x^\vee$, then e does belong to $\text{supp}(\delta_x * \delta_{x^\vee})$.

(SH5) This useful axiom was made by Jewett, but not by Spector. Since it does not hold in all interesting examples, Rösler does not assume this axiom either.

(SH6) Rösler avoids this axiom. Her substitute seems to be the requirement that for each compact subset F of K and each f in $C_c(K)$, the union of all of the supports of all of the translates $y \to f(x * y)$ and $y \to f(y * x)$, as x ranges over F, is precompact.

(SH8) This is a distressing axiom because it is not an axiom for DJS-hypergroups. Moreover, Jewett [J], 5.1D, proves (*) holds provided that the hypergroup has a Haar measure. Of course, many results require the extra assumption that Haar measure exists. While discrete signed hypergroups have a Haar measure using the extra axiom (Pe), without axiom (SH8) it is not known whether compact or commutative hypergroups must have a Haar measure. Parr has examined analogues to axiom (Pe) to see if they might imply the existence of Haar measure. In [Ro], where I set down some tentative axioms for compact signed hypergroups, I avoided (*) by giving an awkward replacement. It has the advantage, though, that I was able to verify it more easily in some examples. The property was created to make the proof of 5.1D in [J] carry over to my setting.

References

[AD] D. Aldous and P. Diaconis, *Shuffling cards and stopping times*, Amer. Math. Monthly **93** (1986), 333–348.

[BD] D. Bayer and P. Diaconis, *Trailing the dovetail shuffle to its lair*, Annals of Applied Prob. **2** (1992), 294-313.

[BH] W. R. Bloom and H. Heyer, *Harmonic Analysis of Probability Measures on Hypergroups*, de Gruyter Studies in Mathematics, 1995.

[BK1] Yu. M. Berezansky and A. A. Kalyushnyi, *Hypercomplex systems and hypergroups: connections and distinctions*, Contemporary Math. AMS, 1995 (Applications of Hypergroups and Related Measure Algebras–Seattle conference, July 31–August 6, 1993), 21–44.

[BK2] Yu. M. Berezansky and Yu. G. Kondratiev, *Biorthogonal systems in hypergroups: an extension of non-Gaussian analysis*, University of Bielefeld preprint.

[CGS] *Applications of Hypergroups and Related Measure Algebras* (Seattle conference, July 31–August 6, 1993), Contemporary Math. AMS, 1995. See also [BK1], [K], [R1] and [Ro].

[D] P. Diaconis, *Group Representations in Probability and Statistics*, Institute of Mathematical Statistics (1988), Hayward, California.

[DH] P. Diaconis and P. Hanlon, *Eigen analysis for some examples of the Metropolis algorithm*, Contemporary Math. **138** (1992), 99–117.

[H] *Probability Measures on Groups and Related Structures*, Eleventh Proceedings- Oberwolfach 1994, edited by H. Heyer, World Scientific Publishing Co. See also [KW] and [R2].

[J] R. I. Jewett, *Spaces with an abstract convolution of measures*, Advances in Math. **18** (1975), 1–101.

[K] T. H. Koornwinder, *Compact quantum Gelfand pairs*, Contemporary Math. AMS, 1995 (Applications of Hypergroups and

Related Measure Algebras–Seattle conference, July 31–August 6, 1993), 213–235.

[KW] V. A. Kaimanovich and W. Woess, *Construction of discrete, non-unimodular hypergroups*, Proc. Oberwolfach 1994 (Probability Measures on Groups and Related Structures), 196–209.

[L] R. Lasser, *Orthogonal polynomials and hypergroups*, Rend. Mat. **3** (1983), 185–209.

[P] A. Parr, *Signed hypergroups*, thesis, University of Oregon, 1997.

[R1] M. Rösler, *Convolution algebras which are not necessarily positivity-preserving*, Contemporary Math. AMS, 1995 (Application of Hypergroups and Related Measure Algebras–Seattle conference, July 31–August 6, 1993), 299–318.

[R2] M. Rösler, *Bessel-type signed hypergroups on* **R**, Proc. Oberwolfach 1994 (Probability Measures on Groups and Related Structures), 292–304.

[R3] M. Rösler, *On the dual of a commutative signed hypergroup*, Manuscr. Math. **88** (1995), 147–163.

[R4] M. Rösler, *Partial characters and signed quotient hypergroups*, preprint.

[Ro] K. A. Ross, *Signed hypergroups–a survey*, Contemporary Math. AMS, 1995 (Applications of Hypergroups and Related Measure Algebras–Seattle conference, July 31–August 6, 1993), 319–329.

[RX1] K. A. Ross and D. Xu, *Hypergroup deformations and Markov chains*, J. Theoretical Probability **7** (1994), 813–830.

[RX2] K. A. Ross and D. Xu, *Metropolis Markov chains on S_n and signed hypergroup deformations*. Unpublished (1993).

[RX3] K. A. Ross and D. Xu, *Some Metropolis Markov chains are random walks on hypergroups*, J. Math. Sciences **28** (1994), 194–234.

[S] R. Spector, *Mesures invariantes sur les hypergroups*, Trans. Amer. Math. Soc. **239** (1978), 147–165.

[W] N. J. Wildberger, *Duality and entropy for finite abelian hypergroups*, preprint, University of New South Wales.

Email: ross@math.uoregon.edu; Fax: (541) 346-0987; Department of Mathematics, University of Oregon, Eugene, OR 97403–1222

Three lectures on Hypergroups Delhi, December 1995

*Alan L. Schwartz**

Abstract

LECTURE 1. Cosines, Legendre polynomials, and Bessel functions are examples of eigenfunctions of Sturm-Liouville problems which are also characters of hypergroups. There are many additional examples from the classical special functions. Indeed, it is possible to give conditions on a Sturm-Liouville problem so that its eigenfunctions must be characters of a hypergroup. The converse will also be discussed. If a hypergroup consists of measures on a real (compact or not) interval, then with adequate regularity conditions, it must be the case that the characters of the hypergroup are eigenfunctions of a Sturm-Liouville problem.

LECTURE 2. A family of orthogonal polynomials on an interval I may be the characters of a hypergroup of measures supported on I (which would be called a continuous polynomial hypergroup), and the family may also supply the characters of a hypergroup of measures on the discrete set $\{0, 1, 2, \dots\}$ (which would be called a discrete polynomial hypergroup). The entire category of continuous polynomial hypergroups can be explicitly described, but the full category of discrete polynomial hypergroups has not yet been characterized, though there are some fairly general theorems.

LECTURE 3. The same issues in the second talk raise analagous questions for multivariate orthogonal polynomials. A family of multivariate orthogonal polynomials is a much more subtle object than a family of one-variable orthogonal polynomials. Some progress has been made in the classification problem, and these results will be discussed as well as recently discovered examples.

---

*The preparation of these lectures took place during the tenure of NSF grant DMS-9404316

Introduction

I have prepared these informal notes to accompany my lectures at the International Conference on Harmonic Analysis, Delhi, December 1995. The purpose of the lectures is to illustrate the interactions among hypergroups, polynomials, and differential equations. The notes are intended as an expanded version of the lectures; neither the notes nor the lectures are offered as a complete survey of the subject, but rather in the hope of whetting the appetite and providing some useful entry points into the literature. I would like to take this opportunity to apologize to the many authors whose work would be included in a more complete survey, but are not included here for lack of space. Some proofs of typical results will be outlined, but the careful reader will have to consult the original memoirs for complete proofs.

The three sections of the notes correspond roughly to the three lectures and provide somewhat more detail than can be covered in an hour.

I have taken the liberty of including many more bibliographic items than are cited in the exposition below in the hope that the reader may find this a convenient list of references to explore the subject.

We mention here three references that are of particular interest to anybody seeking to learn more about this field.

- Dunkl's article [Dun73] is a relatively short exposition in a system somewhat more general than a commutative hypergroup, but it is a nice introduction to the ideas and methods of the subject.

- Jewett's article [Jew75] is a textbook that thoroughly covers the basic properties of hypergroups (although Jewett uses the term "convo" for "hypergroup").

- Litvinov's survey [Lit87], is especially valuable for its broad coverage of much of the work done up to 1985, including such related topics as hypercomplex systems.

- The book of Bloom and Heyer [BH95] gives a unified treatment of the subject including a large body of work on probability in hypergroups; the book includes many examples and a wealth of references.

- The proceedings of a 1993 conference "Applications of Hypergroups and Related Measure Algebras," is published as [CGS95].

Finally, we mention that a number of research questions are gathered in a separate section near the end of these notes.

1. Differential Equations and Hypergroups

As usual, $C(H)$ denotes the continuous functions on H, $M(H)$ the bounded Borel measures on H, and $M_1(H)$ the probability measures on H. The support of a measure μ is denoted $\operatorname{supp}(\mu)$, and the unit point mass at x will be denoted δ_x. H will often be one of the following: $I = [-1,1]$, $J = [0,\pi]$, $\mathbb{R}_+ = [0,\infty)$, $\mathbb{N}_0 = \{0,1,2,\dots\}$.

1.1. Legendre polynomials and beyond

We begin with an old example to illustrate how a family of functions, in this case orthogonal polynomials, can give rise to a measure algebra. I.I. Hirschman, Jr., pointed out in 1956 [Hir56b, Hir56a] that the structure for harmonic analysis exists in a setting where certain orthogonal polynomials could play the role of the exponentials in classical Fourier analysis. For example, consider the Legendre polynomials $\{P_n\}_{n\in\mathbb{N}_0}$. These are orthogonal with respect to Lebesgue measure on I and are normalized by requiring $P_n(1) = 1$. The Legendre polynomials satisfy a product formula:

$$P_n(x)P_n(y) = \int_I K(x,y,z)P_n(z)dz \qquad (-1 < x,\, y < 1)$$

with

$$K(x,y,z) = \begin{cases} \pi^{-1}(1 - x^2 - y^2 - z^2 + 2xyz)^{-1/2}, \\ 0 \end{cases}$$

the first value being taken if and only if $1 - x^2 - y^2 - z^2 + 2xyz > 0$. Obviously $K(x,y,z) \geq 0$ and since $P_0(x) = 1$ it follows from the product formula that

$$\int_I K(x,y,z)dz = 1.$$

For $f,\, g \in L^1 = L^1(I, dx)$ define

$$(f * g)(z) = \int_I \int_I K(x,y,z)f(x)g(y)\,dx\,dy$$

so that

$$\int_I (f*g)(x)P_n(x)\,dx = \left[\int_I f(x)P_n(x)\,dx\right] \cdot \left[\int_I g(x)P_n(x)\,dx\right],$$

and it follows that $(L^1, *)$ is a Banach algebra. The operation is easily extended to the point masses by defining

$$\begin{aligned} &d(\delta_x * \delta_y)(z) = K(x,y,z)\,dz, && (-1 < x,\, y < 1), \\ &\delta_x * \delta_1 = \delta_x \quad \text{and} \quad \delta_x * \delta_{-1} = \delta_{-x}, && (x \in I). \end{aligned}$$

Finally, if μ, $\nu \in M(I)$, define $\mu * \nu$ by its action on an arbitrary continuous function:

$$\int_I f d(\mu * \nu) = \int_I \int_I \int_I f\, d(\delta_x * \delta_y)\, d\mu(x)\, d\nu(y), \qquad (f \in C(H)).$$

Actually, Hirschman discusses these constructions not just for the Legendre polynomials, but for the ultraspherical polynomials $\{P_n^{(\alpha)}(x)\}$ that are orthogonal on I with respect to the measure $(1-x^2)^{\alpha-\frac{1}{2}}\,dx$; the Legendre polynomials are ultraspherical polynomials with $\alpha = 1/2$. The polynomials $R_n^{(\alpha)}(x) = P_n^{(\alpha)}(x)/P_n^{(\alpha)}(1)$ are used in place of the Legendre polynomial $P_n(x)$. In that case

$$K^{(\alpha)}(x,y,z) = \begin{cases} \dfrac{2^{1-2\alpha}(1-x^2-y^2-z^2+2xyz)^{\alpha-1}}{\Gamma^2(\alpha)[(1-x^2)(1-y^2)(1-z^2)]^{\alpha-\frac{1}{2}}}, \\ 0 \end{cases}$$

with the first value being taken if and only if $1-x^2-y^2-z^2+2xyz > 0$.

So for each $\alpha \geq -1/2$ Hirschman obtains a measure algebra that we denote $(I, *_\alpha)$. It is important to note $*_\alpha$ is a distinct convolution for each $\alpha \geq -1/2$, hence a continuum of Banach *algebras* is built on the single Banach *space* $M(I)$. The algebraic structure does not depend on any arithmetic in the underlying space I.

In the special case when $n = 2\alpha + 2$ is an integer, this structure can also be inherited from a group, since $(I, *_\alpha)$ is isometrically isomorphic to the subalgebra of $M(SO(n))$ consisting of measures that are bi-invariant with respect to the action of $SO(n-1)$, or equivalently the measures on the unit sphere in $\mathbb{R}^n$ that are invariant under all rotations that leave some designated point fixed. So $(I, *_\alpha)$ interpolate these in some sense.

For most values of α the group structure is absent, yet there is still an adequate structure in $(I, *_\alpha)$ to define and study objects like Fourier multipliers, maximal functions, and a Littlewood-Paley theory (see [CS77] and the references cited there).

1.2. Hypergroups

$(I, *_\alpha)$ is an example of a hypergroup. Roughly speaking, a hypergroup is a measure algebra that has many of the useful properties associated with the convolution measure algebra of a group, but no presumed algebraic structure for the underlying space. For the sake of clarity we will first give the hypergroup axioms as they apply to the case where the underlying space H is compact; this will be followed by the modifications for the locally compact case. Thus we assume $M(H)$ is a Banach algebra with product $*$. Then $(H, *)$ is a *hypergroup* (or *hypergroup measure algebra*) if it has the following properties:

(H1) If μ and ν are probability measures on H, then so is $\mu * \nu$.

(H2) There is an element $e \in H$ such that $\delta_e * \mu = \mu * \delta_e$ for all $\mu \in M(H)$.

(H3) There is a continuous involution $x \mapsto x^\vee$ ($x^{\vee\vee} = x$) such that $e \in \operatorname{supp} \delta_x * \delta_y$ if and only if $y = x^\vee$.

(H4) $(\mu * \nu)^\vee = \nu^\vee * \mu^\vee$ where $\mu^\vee$ is defined by $\int_H f(x)\, d\mu^\vee(x) = \int_H f(x^\vee)\, d\mu(x)$.

(H5) $(x, y) \mapsto \operatorname{supp}(\delta_x * \delta_y)$ is continuous with an appropriate topology for the space $\mathcal{C}(H)$ of compact subsets of H.

(H6) $(\mu, \nu) \mapsto \mu * \nu$ is weak-* continuous.

The "appropriate topology" in (H5) is the Michael topology [Jew75, §2.5] which has a sub-basis consisting of the sets

$$\mathcal{C}_U(V) = \{K \in \mathcal{C}(H) : K \cap U \neq \emptyset \quad \text{and} \quad K \subset V\}$$

where U and V are arbitrary open subsets of H. If H has a metric ρ, this topology is equivalent to the topology given by the Hausdorff metric on $\mathcal{C}(H)$, which is defined for A, $B \in \mathcal{C}(H)$ by

$$\rho(A, B) = \inf\{r : A \subset V_r(B) \quad \text{and} \quad B \subset V_r(A)\}$$

where $V_r(E) = \{y \in H : \rho(x, y) < r \quad \text{for some} \quad x \in E\}$. A proof of the equivalence is contained in [KS95, Lemma 4.1].

In the locally compact case, we must add to (H5) the requirement that $\operatorname{supp}(\delta_x * \delta_y)$ is compact, and in (H6) "weak-* continuous" must be replaced by "positive continuity," which requires that for each non-negative compactly supported $f \in C(H)$, the mapping $(\mu, \nu) \mapsto \int_H f\, d(\mu * \nu)$ is continuous when restricted to positive measures in $M(H)$.

When H is discrete, for instance $H = \mathbb{N}_0$, (H6) can obviously be dropped and (H5) can be replaced by

(H5) $\operatorname{supp}(\delta_x * \delta_y)$ is finite.

The definitive set of axioms was given first by Jewett in his encyclopaedic article [Jew75]. Jewett calls these objects "convos" and, ironically, never once uses the term "hypergroup."

We also need the following definitions: m is a *Haar measure* for the hypergroup $(H, *)$ if for every $x \in H$, $m * \delta_x = \delta_x * m = m$; m is a *left Haar measure* if at least the second equality holds. $\phi \in C(H)$ is a *character* of $(H, *)$ if ϕ is bounded, $\phi(x^\vee) = \overline{\phi(x)}$ and

$$\int_H \phi \, d(\delta_x * \delta_y) = \phi(x)\phi(y) \qquad (x,\, y \in H).$$

$(H, *)$ is a Hermitian hypergroup if $x^\vee = x$ for every $x \in H$. If $(H, *)$ and $(K, \star)$ are hypergroups, we say they are *equivalent* or identical up to a change of variables if there is a homeomorphism A from H onto K such that if $x = A^{-1}(s)$, and $y = A^{-1}(t)$,

$$\int_K f \, d(\delta_s \star \delta_t) = \int_H (f \circ A) \, d(\delta_x * \delta_y) \qquad (f \in C(K)).$$

When this happens we write $(K, \star) = A(H, *)$. In this case we have $s^\vee = A(x^\vee)$, and ϕ is a character of $(H, *)$ if and only if $\phi \circ A^{-1}$ is a character of $(K, \star)$. If H and K are subsets of vector spaces and A is an affine mapping (i.e., $x \mapsto A(x) - A(0)$ is linear), we say H and K are *linearly equivalent.* Thus, for instance, any hypergroup $(H, *)$ with $H \subset \mathbb{R}^2$ is linearly equivalent to a hypergroup with identity element $(0, 0)$.

The measure algebra of a group with identity e is an obvious example of a hypergroup; convolution is defined by $\delta_x * \delta_y = \delta_{xy}$, and involution by $x^\vee = x^{-1}$, the group inverse of x. In the example, in §1.1 above, $(I, *_\alpha)$ is a Hermitian hypergroup with $e = 1$ and $x^\vee = x$. The characters are $\{R_n^{(\alpha)}(x)\}$ and the orthogonality measure $(1 - x^2)^{\alpha - \frac{1}{2}} \, dx$ is Haar measure. Jewett [Jew75] obtains most of the basic properties necessary to do analysis in hypergroups. In an earlier article Dunkl [Dun73] had shown that a hypergroup $(H, *)$ has a Haar measure if H is compact and $*$ is commutative. Jewett [Jew75] shows that $(H, *)$ has a Haar measure when H is compact, and that it has a left Haar measure when H is discrete. Spector [Spe78] showed $(H, *)$ has a Haar measure when $*$ is commutative. Both Dunkl and Spector use definitions for hypergroup that are slightly different from Jewett's, but their results still apply to the object of Jewett's convos. This is one of the reasons that the object we call a "hypergroup" is now often referred to as a "DJS hypergroup."

1.3. Spherically symmetric random walks

Another early example of a hypergroup arose in probability theory. This example is the subject of a 1963 article by J. F. C. Kingman [Kin63]. Consider a pair of independent random variables $\mathbf{X}$ and $\mathbf{Y}$ in $\mathbb{R}^2$, with lengths X and Y, but with direction uniformly distributed. The sum $\mathbf{Z} = \mathbf{X}+\mathbf{Y}$ also has uniformly distributed direction, but its length $Z = |\mathbf{Z}|$ is a random number in the range $|X-Y| \leq Z \leq X+Y$. In general, if $\mathbf{X}$ and $\mathbf{Y}$ are independent random variables in $\mathbb{R}^2$ with uniformly distributed direction, but with lengths X and Y having probability distributions μ, $\nu \in M_1(\mathbb{R}_+)$, then Z is a random variable in $\mathbb{R}_+$ with a probability distribution depending on μ and ν, denoted by $\mu \circ \nu$, and we write $Z = X \oplus Y$. The operation $\circ$ is readily extended to all of $M(\mathbb{R}_+)$ so that $(M(\mathbb{R}_+), \circ)$ becomes a hypergroup measure algebra that is isometrically isomorphic to the subalgebra of the group convolution algebra $(M(\mathbb{R}^2), *)$ consisting of the measures invariant with respect to rotations of the plane. The characters are indexed by $\mathbb{R}_+$ and given by $\phi_y(x) = J_0(xy)$ where J_α is the Bessel function of the first kind of order α. These satisfy a product formula that yields

$$\int_{\mathbb{R}_+} \phi_y \, d(\mu * \nu) = \left[\int_{\mathbb{R}_+} \phi_y \, d\mu \right] \cdot \left[\int_{\mathbb{R}_+} \phi_y \, d\nu \right] ,$$

so that the useful substitute for the characteristic function of the random variable X is $\Phi_X(y) = \int_{\mathbb{R}_+} \phi_y \, d\mu$. The product formula for the Bessel functions also ensures the fundamental property of characteristic equations $\Phi_{X \oplus Y} = \Phi_X \Phi_Y$ when X and Y are independent random variables in $\mathbb{R}_+$.

Kingman actually describes a continuum of Hermitian hypergroups $(\mathbb{R}_+, \circ_\alpha)$ (of course, he never uses the word "hypergroup"). The identity element is 0, and the characters are given by $\phi_y(x) = \mathcal{J}_\alpha(yx) = 2^\alpha \Gamma(\alpha + 1)(yx)^{-\alpha} J_\alpha(yx)$ for $y \in \mathbb{R}_+$.

When $n = 2\alpha + 2$ is an integer, $(\mathbb{R}_+, \circ_\alpha)$ is isometrically isomorphic to the subalgebras of rotation invariant measures on $\mathbb{R}^n$. There is again no useful algebraic structure in the underlying spaces. Nevertheless Kingman is able to define random walk and Brownian motion, and obtain a law of large numbers, a central limit theorem, a recurrence theorem, and characterizations of infinitely divisible and stable distributions. When $n = 2\alpha + 2$ is an integer, all of this is an inheritance from the group structure on $\mathbb{R}^n$, but Kingman obtains his results for all real $\alpha \geq -1/2$ with no reference to the group case except for inspiration.

1.4. Differential equations

The examples discussed above have the property in common that the characters are eigenfunctions of a second-order linear differential operator. In the case of the ultraspherical example, it is convenient to make the change of variables $x = \cos\theta$ and write $u_n^{(\alpha)}(\theta) = R_n^{(\alpha)}(\cos\theta)$. These functions (for $n \in \mathbb{N}_0$) constitute a complete set of eigenfunctions for a Sturm-Liouville problem on $[0, \pi]$:

$$(\rho^2 y')' + \lambda\rho^2 y = 0 \qquad y'(0) = y'(\pi) = 0\ , \tag{1.1}$$

with $\rho(\theta) = (\sin\theta)^\alpha$. The modified Bessel functions $\mathcal{J}_\alpha(yx)$ are the eigenfunctions for the following Sturm-Liouville problem on $\mathbb{R}_+$:

$$(Ay')' + \lambda Ay = 0 \qquad y'(0) = 0 \tag{1.2}$$

with $A(x) = x^{2\alpha+1}$. This leads us to ask about the following:

Sufficient conditions. Which Sturm-Liouville problems have eigenfunctions that are characters of a hypergroup? In such cases, the full structure of hypergroups is available to elucidate expansions in terms of the eigenfunctions.

Necessary conditions. Which hypergroups on real intervals have characters that are eigenfunctions of Sturm-Liouville problems? In such cases, the differential equation yields detailed information about the characters.

1.5. Sufficient conditions

The earliest relevant result we know of is Chébli [Ché72]. He considers the differential equation (1.2) with A satisfying the following conditions:

1. $A(0) = 0$.
2. $A(x) > 0$ for $x > 0$.
3. $A'(x)/A(x) = \alpha x^{-1} + B(x)$ where B is continuous at 0.
4. $A(x)$ increases to ∞ as x increases to ∞.

This leads to the following result:

Theorem 1.1. *Let $a(x, y) = A(x)A(y)$. Suppose f is positive and twice continuously differentiable on $\mathbb{R}_+$ and $u(x, y)$ satisfies*

$$(au_x)_x = (au_y)_y, \qquad u(x, 0) = f(x), \qquad u_y(x, 0) = 0.$$

Then $u(x, y) \geq 0$.

This gives rise to a hypergroup on $\mathbb{R}_+$ with characters that satisfy (1.2) by a route similar to that which will be outlined below for the compact case.

Since 1972 a variety of similar results have emerged as well as many in the compact case; several of the latter are discussed and compared in [CS90b]. We outline the path given in [CS90b], which leads from a general Sturm-Liouville problem (1.1) to the corresponding hypergroup. We begin with some assumptions on ρ.

1. ρ is positive and continuous on $(0, \pi)$.

2. $\rho(\pi - s) = \rho(s)$.

3. $\rho'(s)/\rho(s)$ is non-increasing in $(0, \pi)$.

4. $\rho(s) = (\sin s)^\gamma g(s)$ for some $\gamma \geq 0$. g must be real-analytic at 0 and have p continuous derivatives on $(0, \pi)$ where $p \geq \max(\gamma + \frac{1}{2}, 2)$.

It follows that the eigenvalues of (1.1) form an increasing sequence $0 = \mu_0 < \mu_1 < \mu_2 < \ldots$. Let y_k be the eigenfunction corresponding to μ_k normalized to $y_k(0) = 1$ so that, in particular, $\phi_0(s) = 1$. Then

$$y_k(\pi - s) = (-1)^k y_k(s) \ , \tag{1.3}$$

and $\{y_k\}$ is a complete orthogonal family for $L^2([0, \pi], \rho^2(s)\, ds)$. The key fact is that $\{y_k\}$ satisfies a product formula:

Theorem 1.2. *For each s, $t \in J = [0, \pi]$ there is $\sigma_{s,t} \in M_1(J)$ such that*

1. $\int_J y_k \, d\sigma_{s,t} = y_k(s) y_k(t) \qquad (k \in \mathbb{N}_0)$.

2. $\operatorname{supp} \sigma_{s,t} \subset [|s - t|, \, \pi - |s + t - \pi|]$.

3. $f(s, t) = \int_J f \, d\sigma_{s,t}$ *is continuous on* $J \times J$.

Thus there is a Hermitian hypergroup $(J, *)$ with character set $\{y_k\}$, Haar measure $\rho^2(s)\, ds$, $e = 0$ and convolution of two meaures μ, $\nu \in M(J)$ defined by their action on $f \in C(J)$:

$$\int_J f \, d(\mu * \nu) = \int_J \int_J \left[\int_J f \, d\sigma_{s,t} \right] d\mu(s)\, d\nu(t).$$

In particular, $\delta_s * \delta_t = \sigma_{s,t}$.

The proof of the theorem is based on two lemmas. Let $\mathcal{P}$ be the trigonometric polynomials, that is, the space of finite linear combinations from $\{y_k\}$.

Lemma 1.3. *$\mathcal{P}$ is dense in $C(J)$ in the topology of uniform convergence.*

Proof. The differential equation can be used to show that if f has $2p$ continuous derivatives and compact support in $(0, \pi)$, then the Fourier series of f with respect to $\{y_k\}$ converges absolutely and uniformly to f. ∎

Lemma 1.4. *If*

$$f(s) = \sum_{k=0}^{n} c_k y_k(s) \geq 0, \qquad (s \in J) \tag{1.4}$$

then

$$f(s,t) = \sum_{k=0}^{n} c_k y_k(s) y_k(t) \geq 0, \qquad ((s,t) \in J \times J). \tag{1.5}$$

Proof. For $0 \leq a \leq b \leq \pi/2$ let $\Delta(a,b)$ be the triangle with vertices (a,b), $(a-b,0)$, and $(a+b,0)$. We begin by assuming that (1.4) holds with $>$ in place of $\geq$; we will show that (1.5) holds in $\Delta(\pi/2,\pi/2)$. Assume by way of contradiction that (1.5) fails at some point of $\Delta(\pi/2,\pi/2)$. Then it is possible to select $P = (\xi,\eta) \in \Delta(\pi/2,\pi/2)$ such that $f(\xi,\eta) = 0$ but $f(s,t) > 0$ if $(s,t) \in \Delta(\xi,\eta) - \{P\}$.

The rest of the argument is built around a related hyperbolic Cauchy problem (as in Chébli's theorem). Let $W(s,t) = \rho^2(s)\rho^2(t)$. $f(s,t)$ satisfies

$$(Wf_s)_s - (Wf_t)_t = 0, \qquad f(s,0) = f(s), \qquad f_t(s,0) = 0.$$

Now let $C = (\xi - \eta, 0)$ and $D = (\xi + \eta, 0)$ then Green's Theorem yields

$$\begin{aligned} 0 &= \iint_{\Delta(\xi,\eta)} \Big[(Wf_s)_s - (Wf_t)_t\Big]\, ds\, dt \\ &= \int_{\partial\Delta(\xi,\eta)} (Wf_s\, dt + Wf_t\, dt) = -\Big[\int_{CP} + \int_{DP}\Big] W\, df. \end{aligned}$$

Integration by parts yields

$$\begin{aligned} 2W(P)f(P) &= W(C)f(C) + W(D)f(D) \\ &\qquad + \int_{CP} f(W_t + W_s)\, dt + \int_{DP} f(W_t - W_s)\, dt\ , \end{aligned}$$

which implies $f(P) > 0$. This contradicts the original assumption.

Now if we simply assume (1.4) and let ϵ be any positive number, then

$$f_\epsilon = f + \epsilon = f + \epsilon y_0 > 0,$$

thus for any $(s,t) \in \Delta(\pi/2, \pi/2)$,

$$f_\epsilon(s,t) = f(s,t) + \epsilon y_0(s)y_0(t) = f(s,t) + \epsilon > 0,$$

whence $f(s,t) \geq 0$ for $(s,t) \in \Delta(\pi/2, \pi/2)$. Equation (1.5) is now obtained on all of $J \times J$ by employing the symmetries $f(t,s) = f(s,t)$ and $f(\pi - s, \pi - t) = f(s,t)$ (recall (1.3)). ∎

Proof of theorem. If f is a non-negative function in $C(J)$, then $0 \leq f(s) \leq \|f\|_\infty$, so Lemma 1.4 shows $0 \leq u(s,t) \leq \|f\|_\infty$. Thus $\sigma_{s,t} \in M_1(J)$. The other properties are all readily verified. ∎

Other results can be obtained by exploiting the maximum principal for hyperbolic equations or using a method devised by C. Markett [Mar89, CMS91, CMS92, CMS93] based on the Riemann integration method for such equations. This latter method has the advantage of yielding an explicit description of $\sigma_{s,t}$ and showing that it is absolutely continuous for $0 < s,\ t < \pi$. That method yields product formulas (and the associated hypergroups) on an interval with modified Bessel functions [Mar89] and spheroidal wave functions [CMS93] as characters.

1.6. Necessary conditions

The question here is when a one-dimensional hypergroup $(H, *)$ (meaning one in which H is homeomorphic to the circle or a real interval) has characters tjat are eigenfunctions of a Sturm-Liouville problem. The first piece of the problem is to obtain some information about one-dimensional hypergroups. All such hypergroups are commutative, and there are essentially four types [Sch88, Zeu89]:

1. $H = \mathbb{R}$ and $*$ is classical convolution. In this case $e = 0$ and $x^\vee = -x$.

2. $H = \mathbb{T}$ (where $\mathbb{T}$ is the unit circle in the complex plane) and $*$ is the classical convolution. In this case $e = 1$ and $z^\vee = \overline{z}$.

3. H is a compact interval, e is one endpoint, and $x^\vee = x$.

4. H is a half-open interval (not necessarily bounded), e is the endpoint, and $x^\vee = x$.

1.7. Jacobi type hypergroups

We will concentrate on the situation where H is a compact interval; indeed, it is no loss of generality to require $H = J = [0, \pi]$. We will give sufficient conditions on a hypergroup $(J, *)$ so that its characters are exactly the eigenfunctions of a Sturm-Liouville problem (1.1). With each such hypergroup we will associate a pair of parameters (α, β) that can be used to help describe the properties of the hypergroup. The definition is somewhat complicated, but it includes (up to a change of variables) every hypergroup $(H, *)$, of which we are aware, where H is a compact interval. The discussion below is outlined from [CS90a].

We begin by requiring that $(J, *)$ satisfies the following differentiablity condition: If f has p continuous derivatives and is compactly supported in $(0, \pi)$, then

$$f(s,t) = \int_J f \, d(\delta_s * \delta_t)$$

has bounded pth-order derivatives on the interior of $J \times J$. For $\mu = 1$, 2 let k_μ be the largest positive integer such that

$$M_\mu(s,t) = \int_J (r-t)^\mu \, d(\delta_s * \delta_t)(r) = O(s^{k_\mu}).$$

When k_μ is finite,

$$M_\mu(s,t) = A_\mu(t) s^{k_\mu} + o(s^{k_\mu}).$$

$(J, *)$ is a *Jacobi type* (α, β) *hypergroup* if

1. $k_1 = k_2 = 2$.
2. $(\sin t) A_1(t)$ is differentiable on J.
3. $A_2(t)$ is a positive constant a_2.
4. α and β satisfy

$$\lim_{t \to 0+} \frac{\sin t}{a_2} A_1(t) = \alpha + \frac{1}{2} \qquad \text{and} \qquad \lim_{t \to \pi-} \frac{\sin t}{a_2} A_1(t) = -(\beta + \frac{1}{2}) \, .$$

The conditions are not strict as they seem, since a much larger class of hypergroups are equivalent to Jacobi type hypergroups by a change of variable, so that it is really only necessary that k_1 be finite, that $(\sin t) A_1(t)$ and $A_2(t)$ can be extended to be positive differentiable functions on J, and that $A_2(t)$ be positive.

The definition is motivated by the example of the Jacobi polynomials $P_n^{(\alpha,\beta)}$, which are orthogonal on I with respect to the weight $(1-x)^\alpha(1+x)^\beta$. These include the ultraspherical polynomials since $R_n^{(\alpha)} = R_n^{(\alpha-\frac{1}{2},\alpha-\frac{1}{2})}$. For many values of (α,β) the Jacobi polynomials normalized to the value 1 at $x = 1$ are the characters of a hypergroup (which will be discussed in the next lecture). That hypergroup with the change of variables $x = \cos\theta$ satisfies (1)–(5). The functions $\phi_n(t) = P_n^{(\alpha,\beta)}(\cos t)/P_n^{(\alpha,\beta)}(1)$ are the eigenfunctions of the Sturm-Liouville problem (1.1) for $\rho = (\sin t/2)^{\alpha+\frac{1}{2}}(\cos t/2)^{\beta+\frac{1}{2}}$.

Now assume $(J,*)$ is a Jacobi type (α,β) hypergroup. Let

$$\rho(t) = c\exp\left(\int_{\pi/2}^{t}\frac{A_1(r)}{a_2}\,dr\right),$$

where c is chosen so that $\int_0^\pi \rho^2(t)\,dt = 1$. The eigenvalues of (1.1) can be arranged in an increasing sequence $\lambda_0 < \lambda_1 < \lambda_2 < \ldots$; let ϕ_k be the eigenfunction corresponding to λ_k and satisfying $\phi_k(0) = 1$; these are the characters of $(J,*)$. We assume α, $\beta \geq -1/2$. Then it is possible to obtain some precise estimates on the characters. First we introduce several constants. There are positive constants a_α and a_β such that

$$\lim_{t\to 0+}(\sin t)^{-\alpha-\frac{1}{2}}\rho(t) = a_\alpha, \qquad \lim_{t\to\pi-}(\sin t)^{-\beta-\frac{1}{2}}\rho(t) = a_\beta.$$

Let $E_\alpha = 2^\alpha\Gamma(\alpha+1)a_\alpha$, $E_\beta = 2^\beta\Gamma(\beta+1)a_\beta$, and $E = E_\alpha/E_\beta$. Then it is possible to obtain the following information about $(J,*)$:

1. $\rho^2(t)\,dt$ is Haar measure for $(J,*)$.

2. $\alpha \geq \beta \geq -1/2$.

3. $\lim_{k\to\infty}(-1)^k k^{\alpha-\beta}\phi_k(\pi) = E$.

4. $\lim_{k\to\infty}\int_J \phi_k^2(t)\,dt = E_\alpha^2$.

5. $\lim_{k\to\infty}\lambda_k/k = 1$.

6. There is $K > 0$ such that $\phi_k(t) \geq \frac{1}{3}$ provided $kt \leq K$.

7. There are constants C and C_ϵ for any $\epsilon > 0$ such that

$$\phi_k(s) = a_\alpha\frac{s^{\alpha+\frac{1}{2}}}{\rho(s)}\mathcal{J}_\alpha(\lambda_k s) + I_k(s)$$

where

$$|I_k(s)| = \begin{cases} Ck^{-1} & 0 \leq s \leq 1/k \\ C_\epsilon \dfrac{\ln k}{k(sk)^{\alpha+\frac{1}{2}}} & k^{-1} \leq s \leq \pi - \epsilon. \end{cases}$$

There is a similar estimate at π.

These properties facilitate a study of the harmonic analysis of $(J, *)$, so that one can obtain, for instance, a useful analogue of the Hardy-Littlewood maximal inequality [CS89].

2. Orthogonal Polynomials and Hypergroups

2.1. Ultraspherical polynomials, again

Hirschman [Hir56b, Hir56a] also discusses an operation on $\ell^1 = M(\mathbb{N}_0)$ based on the linearization formula for ultraspherical polynomials:

$$R_n^{(\alpha)}(x) R_m^{(\alpha)}(x) = \sum_{k=|n-m|}^{n+m} c_{n,m}^k R_k^{(\alpha)}(x). \tag{2.1}$$

The explicit formula quoted by Hirschman shows $c_{n,m}^k \geq 0$, and if we set $x = 1$ in (2.1) we obtain $\sum_{k \in \mathbb{N}_0} c_{n,m}^k = 1$. Thus if a, $b \in \ell^1 = M(\mathbb{N}_0)$ we can define their convolution $a \star_\alpha b$ by

$$(a \star_\alpha b)_k = \sum_{n=0}^{\infty} \sum_{m=0}^{\infty} c_{n,m}^k a_n b_m$$

or equivalently for each $x \in I$,

$$\sum_{k \in \mathbb{N}_0} (a \star_\alpha b)_k R_k^{(\alpha)}(x) = \left[\sum_{n \in \mathbb{N}_0} a_n R_n^{(\alpha)}(x) \right] \cdot \left[\sum_{m \in \mathbb{N}_0} b_m R_m^{(\alpha)}(x) \right].$$

It is easy to check that $M(\mathbb{N}_0)$ with this operation is a hypergroup which we denote by $(\mathbb{N}_0, \star_\alpha)$. Thus Hirschman describes two hypergroups $(I, *_\alpha)$ and $(\mathbb{N}_0, \star_\alpha)$ that have many properties analogous to the classical convolution algebras of measures on the circle group and its dual on the group of integers.

It is important to note $\star_\alpha$ is a distinct convolution for each $\alpha \geq -1/2$, hence a continuum of Banach *algebras* is again built on the single Banach *space* $M(\mathbb{N}_0)$. The algebraic structure does not depend on any arithmetic in the underlying space $\mathbb{N}_0$.

2.2. Orthogonal polynomials

It is possible to formulate linearization and product formulas for general families of orthogonal polynomials and perhaps to imitate the construction of the two measure algebras as in the ultraspherical case. But there can be no guarantee in general that these operations are truly well-defined. We are interested in when the constructions actually yield hypergroups.

Thus suppose $m \in M_1(\mathbb{R})$; let $H = \operatorname{supp} m$. $\{P_n\}$ is a family of orthogonal polynomials with respect to m if for each n, P_n is a polynomial of degree n that satisfies

$$\int_H x^m P_n(x)\, dm(x) = 0 \qquad (m = 0, 1, \ldots, n-1).$$

These conditions determine P_n up to a non-zero constant. If x_0 is chosen so that none of the polynomials vanishes at x_0, we can define the normalized polynomials

$$R_n(x) = P_n(x)/P_n(x_0) \qquad \text{so that} \qquad R_n(x_0) = 1.$$

A family of orthogonal polynomials always has a *linearization formula*

$$R_n R_m = \sum_{k=0}^{n+m} c_{n,m}^k R_k,$$

where

$$c_{n,m}^k = \frac{\int_H R_k R_n R_m\, dm}{\int_I R_k^2\, dm}.$$

Clearly, the orthonormality of $\{R_n\}$ implies $c_{n,m}^k = 0$ if $k > n+m$ or $n > m+k$ or $m > n+k$; that is, the linearization formula can be rewritten

$$R_n R_m = \sum_{k=|n-m|}^{n+m} c_{n,m}^k R_k. \tag{2.2}$$

Evaluation of (2.2) at $x = x_0$ yields

$$\sum_{k=|n-m|}^{n+m} c_{n,m}^k = 1. \tag{2.3}$$

If in fact

$$c_{n,m}^k \geq 0\ , \tag{2.4}$$

it is possible to define the product of two sequences a, $b \in M(\mathbb{N}_0) = \ell^1$ by setting

$$(a \star b)_k = \sum_{n=0}^{\infty} \sum_{m=0}^{\infty} c_{n,m}^k a_n b_m.$$

This is equivalent to saying

$$a \star b = c \Longleftrightarrow \left[\sum_{n=0}^{\infty} a_n R_n(x)\right] \cdot \left[\sum_{m=0}^{\infty} b_m R_m(x)\right] = \left[\sum_{k=0}^{\infty} c_k R_k(x)\right],$$

at least for finitely supported sequences a and b. Equations (2.3) and (2.4) guarantee that $M(\mathbb{N}_0)$ is a Banach algebra. In fact, $(\mathbb{N}_0, \star)$ is a Hermitian hypergroup with identity element $e = 0$ and each character has the form

$$\phi_z(a) = \sum_{n=0}^{\infty} a_n R_n(z)$$

provided $z \in \mathbb{C}$ satisfies $\operatorname{lub} |R_n(z)| < \infty$. Such a hypergroup is called the *discrete polynomial hypergroup associated with* $\{R_n\}$.

A family of orthogonal polynomials always has a linearization formula, but there is no guarantee that it will have a product formula, which is the continuous analog of the linearization formula. That is, we say $\{R_n\}$ has a *product formula* if for each pair x, $y \in H$, there is a fixed measure $\sigma_{x,y} \in M(H)$ such that for every $n \in \mathbb{N}_0$

$$R_n(x) R_n(y) = \int_H R_n \, d\sigma_{x,y}. \tag{2.5}$$

If $\|\sigma_{x,y}\|$ are uniformly bounded, it is possible to define a continuous product $*$ on $M(H)$ by the formula

$$\int_H f d(\mu * \nu) = \int_H \int_H \int_H f \, d\sigma_{x,y} \, d\mu(x) \, d\nu(y),$$

so that, in particular, $\delta_x * \delta_y = \sigma_{x,y}$. If $\sigma_{x,y}$ is a non-negative measure, evaluation of (2.5) for $n = 0$ shows that $\sigma_{x,y} \in M_1(H)$ so that $M(H)$ becomes a Banach algebra with operation $*$. It also follows from (2.5) evaluated at $x = x_0$ that $\sigma_{x_0,y} = \delta_y$, so that if in fact this Banach algebra is a hypergroup, it is Hermitian with identity element $e = x_0$. In that case we say that $\{R_n\}$ has a *hypergroup product formula* and we call $(H, *)$ the *continuous polynomial hypergroup* associated with $\{R_n\}$.

When a family of orthogonal polynomials has both a continuous and a discrete hypergroup associated with it, the two hypergroups are in some sense dual to one another in analogy with the way that the circle and the integers are a pair of dual groups.

We ask the following questions:

(Q1) Which hypergroups $(\mathbb{N}_0, \star)$ are in fact discrete polynomial hypergroups?

(Q2) Which families of orthogonal polynomials give rise to discrete polynomial hypergroups?

and the corresponding questions on the continuous side:

(Q3) Which hypergroups $(H, *)$ with $H \in \mathbb{R}$ are in fact continuous polynomial hypergroups?

(Q4) Which families of orthogonal polynomials give rise to continuous polynomial hypergroups?

2.3. Jacobi polynomials

The normalized Jacobi polynomials $R_n^{(\alpha,\beta)}$ have product and linearization formulas for many values of the parameters. The region in the (α, β)-plane where this happens was identified by Gasper [Gas70, Gas72], consequently there is a continuous polynomial hypergroup $J(\alpha, \beta) = (I, *_{(\alpha,\beta)})$ associated with $\{R_n^{(\alpha,\beta)}\}$ if and only if (α, β) belongs to the set

$$E_J = \{(\alpha, \beta) : \alpha \geq \beta > -1 \quad \text{and} \quad \beta \geq -1/2 \quad \text{or} \quad \alpha + \beta \geq 0\}$$

and there is a discrete polynomial hypergroup $(\mathbb{N}_0, \star_{(\alpha,\beta)})$ associated with $\{R_n^{(\alpha,\beta)}\}$ if and only if (α, β) belongs to the set

$$F_J = \{(\alpha, \beta) : a(a+3)^2(a+5) \geq (a^2 - 7a - 24)b^2 \quad \text{where} \quad a = \alpha + \beta + 1 \quad \text{and} \quad b = \alpha - \beta\}.$$

The set F_J contains E_J as a proper subset, so there can be no hope of an analog of Pontryagin duality for hypergroups. These include the examples discussed by Hirschman since the continuous polynomial hypergroup associated with $\{P_n^{(\alpha)}\}$ is simply $J(\alpha - \frac{1}{2}, \alpha - \frac{1}{2})$.

2.4. Discrete polynomial hypergroups

We begin with a few observations. If ϕ is a character of a hypergroup $(H, *)$ then $\|\phi\|_\infty = \sup_{x \in H} |\phi(x)| = 1$. To see this, observe that $\phi(x)\phi(y) = \int_H \phi \, d(\delta_x * \delta_y)$ immediately leads to $\|\phi\|_\infty^2 \leq \|\phi\|_\infty$, which implies $\|\phi\|_\infty \leq 1$. On the other hand, $\|\phi\|_\infty \geq \phi(e) = 1$, if $(H, *)$ is a discrete polynomial hypergroup associated with a family of orthogonal polynomials $\{R_n\}$. Since for $n > 0$, R_n is an unbounded continuous function, it follows that H must be a bounded set. A simple change of variables allows us to insist that $H \subset I$ with $e = 1$. H is necessarily Hermitian and commutative.

We recall that every family of orthogonal polynomials satisfies a recurrence relation

$$xR_n(x) = \gamma_n R_{n+1}(x) + \beta_n R_n(x) + \alpha_n R_{n-1}(x) \tag{2.6}$$

and there are sufficient conditions [Fav35, Sho36] that polynomials that satisfy such a relation are orthogonal with respect to a compactly supported non-negative measure. (Q1) has a fairly straightforward answer based on this last observation [Sch77].

Theorem 2.1. $(\mathbb{N}_0, \star)$ *is the discrete polynomial hypergroup associated with a family* $\{R_n\}$ *of orthogonal polynomials if and only if*

1. $(\delta_n \star \delta_1)(\{k\}) = 0$ *if* $|n-k| > 1$.
2. $(\delta_n \star \delta_1)(\{n+1\}) > 0$.

Moreover, the polynomials may be normalized so that $R_1(x) = x$ *and the orthogonality measure is supported in* I.

There are two types of responses to the harder question (Q2). The first is a large collection of results such as Gasper's that have been obtained by workers in special functions. There are several general results based on the recurrence relation (2.6). The first result we know of is due to R. Askey [Ask70], but it is subsumed by results due to Szwarc [Szw92a, Szw92b]; we state one from [Szw92a], but he has proved other theorems that are more widely applicable.

Theorem 2.2. *If polynomials* P_n *satisfy*

$$xP_n = \gamma_n P_{n+1} + \beta_n P_n + \alpha_n P_{n-1}$$

and

(i) α_n, β_n *and* $\alpha_n + \gamma_n$ *are increasing sequences* (γ_n, $\alpha_n \geq 0$),

(ii) $\alpha_n \leq \gamma_n$ *for* $n \in \mathbb{N}_0$,

then

$$P_n P_m = \sum_{k=|n-m|}^{n+m} c_{n,m}^k P_k$$

with $c_{n,m}^k \geq 0$.

This really is a discrete analogue to the Sturm-Liouville approach of the first lecture, since (2.6) may be reformulated as an eigenvalue difference relation:

$$\gamma_n \Delta^2 R_n(x) + (\gamma_n - \alpha_n)\Delta R_n(x) + (\gamma_n + \beta_n) R_n(x) = x R_n(x) \ ,$$

where $\Delta^2 R_n(x) = R_{n+1}(x) - 2R_n(x) + R_{n-1}(x)$ and $\Delta R_n(x) = R_n(x) - R_{n-1}(x)$. In fact, Szwarc's method is based on a discrete analogue of the hyperbolic Cauchy problem described in the first lecture. Markett has also devised a method generalizing the Riemann integration technique, which he can also use to obtain explicit linearization formulas [Mar94].

We refer the reader to the recent volume by Bloom and Heyer [BH95], which lists many more examples of discrete polynomial hypergroups (they use the term "polynomial hypergroups" for discrete polynomial hypergroups).

2.5. Continuous polynomial hypergroups

The story here is a much simpler one and can be found in [CS90c, CMS92, CS95b]; complete answers to (Q3) and (Q4) are contained in the following theorems.

Theorem 2.3. *If H is an interval then $(H, *)$ is a continuous polynomial hypergroup if and only if $(H, *)$ is equivalent by means of a linear change of variables to a continuous Jacobi polynomial hypergroup $J(\alpha, \beta)$ with $(\alpha, \beta) \in E_J$.*

Outline of proof. The "if" is essentially Gasper's result (see §). The "only if" is based on the idea of obtaining a second order linear differential equation for the characters, and then using an old result of Bochner [Boc29], which shows that there are only a few families of polynomials which satisfy such an equation. So we assume that $(H, *)$ is a continuous polynomial hypergroup associated with an orthogonal family $\{R_n\}$. To obtain the differential equation, begin with the relation

$$R_n(s)R_n(t) = \int_H R_n(r)\, d(\delta_s * \delta_t)(r).$$

We now obtain a differential equation for R_n by following a path laid out in [Lev64]. Expansion of $R_n(s)$ in a Taylor series around $s = e$ and of $R_n(r)$ around $r = t$ yields

$$\sum_{k=0}^{n} \frac{1}{k!} R_n(e)(s-e)^k R_n(t) = \sum_{k=0}^{n} \frac{1}{k!} M_k(s,t) R_n(t) \qquad (2.7)$$

where $M_k(s,t) = \int_H (r-t)^k d(\delta_s * \delta_t)(r)$. $M_1(s,t)$ and $M_2(s,t)$ can be computed explicitly in terms of R_1 and R_2 and then expressed in terms of $s-e$ and $(s-e)^2$. For $k > 0$, $M_k(s,t)$ has no $s-e$ terms since $M_k(s,t) = o(M_2(s,t))$ if $k > 2$. The coefficients of $s-e$ on both sides of (2.7) are then equated to obtain

$$A(t)y''(t) + B(t)y'(t) = \lambda_n y(t) \qquad (y = R_n, \quad \lambda_n = R_n'(e)).$$

■

The same proof can be used to prove a result that weakens the condition that H be an interval:

Theorem 2.4. *Suppose that $(H,*)$ is a hypergroup where H is an infinite subset of $\mathbb{R}$ and that for every $n \in \mathbb{N}_0$ $(H,*)$ has a character which is a polynomial R_n of degree n. Assume one of the following holds:*

1. *H is compact*
2. *$\{R_n\}$ is orthogonal with respect to a positive Borel measure on H.*

Then $(H,*)$ is equivalent by means of a linear change of variables to a continuous Jacobi polynomial hypergroup $J(\alpha,\beta)$ with $(\alpha,\beta) \in E_J$.

3. Multivariate orthogonal polynomials and hypergroups

3.1. Multivariate orthogonal polynomials

The notion of orthogonal polynomials is somewhat more complicated for multivariate polynomials than it is for polynomials of one variable. In the latter case, the orthogonality measure determines the polynomials up to a constant multiple. In the multivariate case, the measure only determines a subspace $\mathcal{V}_n$ of polynomials (see below).

We first need some notation. We use following: Let $\nu \in \mathbb{N} = \{1,2,3,\dots\}$ and let

$$\begin{aligned}
\mathbf{x} &= (x_1, x_2, \dots, x_\nu), \qquad (x_1, x_2, \dots, x_\nu \in \mathbb{R}) \\
d\mathbf{x} &= dx_1\, dx_2\, \dots dx_\nu, \\
\mathbf{k} &= (k_1, k_2, \dots, k_\nu), \qquad (k_1, k_2, \dots, k_\nu \in \mathbb{N}_0), \\
|\mathbf{k}| &= k_1 + k_2 + \dots + k_\nu, \\
\mathbf{x}^{\mathbf{k}} &= x_1^{k_1} x_2^{k_2} \dots x_\nu^{k_\nu}.
\end{aligned}$$

If p is a polynomial then $\deg p$ is the highest degree of a monomial that occurs in p with a non-zero coefficient. Let $m \in M(\mathbb{R}^\nu)$ and let $H = \operatorname{supp} m$. Then m uniquely determines the subspaces

$$\mathcal{V}_n = \{p : \deg p = n \quad \text{and} \quad \int_H \mathbf{x}^\mathbf{k} p(\mathbf{x})\, dm(\mathbf{x}) = 0 \quad \text{if} \quad |\mathbf{k}| < n\}.$$

When $\nu = 1$, $\dim \mathcal{V}_n = 1$ because the orthogonal polynomials in one variable are determined up to a multiplicative constant. In the general case $\dim \mathcal{V}_n > 1$; for instance when $\nu = 2$, $\dim \mathcal{V}_n = n + 1$. Thus the measure does not determine the individual orthogonal polynomials. One could construct the orthogonal polynomials of each degree by first selecting an order for the set of monomials $\{\mathbf{x}^\mathbf{k} : |\mathbf{k}| = n\}$, and then applying the Gram–Schmidt orthogonalization process. But there are infinitely many possibilities not included among these $n!$ choices, since the problem is really that of choosing an orthogonal basis for the vector space $\mathcal{V}_n$ with respect to the inner product $\langle f, g\rangle = \int_H f(\mathbf{x})\overline{g}(\mathbf{x})\, dm(\mathbf{x})$. We say a set $\mathcal{P}$ of ν-variable polynomials is *algebraically complete* if $\mathcal{P}$ is linearly independent and if for each $n \in \mathbb{N}_0$, $\{p \in \mathcal{P} : \deg p \leq n\}$ spans the space of all polynomials of degree not exceeding n. When $\nu = 1$ an algebraically complete family is simply a set of polynomials that contains exactly one of each degree. We can obtain an algebraically complete family of polynomials by choosing a basis for each $\mathcal{V}_n$ and letting $\mathcal{P}$ be the union of those bases. In this case we say that $\mathcal{P}$ is a *family of polynomials orthogonal with respect to m.*

3.2. Continuous and discrete multivariate polynomial hypergroups

Suppose that $(H, *)$ is a hypergroup with $H \subset \mathbb{R}^\nu$. We say that $(H, *)$ is a *continuous ν-variable polynomial hypergroup* if the character set of $(H, *)$ contains a ν-variable algebraically complete family of polynomials. A definition of discrete ν-variable polynomial hypergroup can be readily formulated, but will not be needed in this lecture. The next few sections contain examples of multivariate continuous polynomial hypergroups.

3.3. Example. Product hypergroups; hypergroups on a square

It is always possible to obtain a hypergroup as a direct product of two hypergroups and in this way obtain continuous ν-variable polynomial hypergroups for every $\nu \in \mathbb{N}$. In particular, if $(H_1, *_1)$ and $(H_2, *_2)$ are hypergroups, then so is

$$(H, *) = (H_1, *_1) \otimes (H_2, *_2)$$

where $H = H_1 \times H_2$. The product of point masses is given by the product measure

$$\delta_{(x_1,x_2)} * \delta_{(y_1,y_2)} = (\delta_{x_1} *_1 \delta_{y_1}) \times (\delta_{x_2} *_2 \delta_{y_2}).$$

Thus if $(\boldsymbol{\alpha}, \boldsymbol{\beta}) = (\alpha_1, \beta_1; \alpha_2, \beta_2) \in E_J \times E_J$, $J(\boldsymbol{\alpha}, \boldsymbol{\beta}) = J(\alpha_1, \beta_1) \otimes J(\alpha_2, \beta_2)$ is an example of a Hermitian continuous 2-variable polynomial hypergroup with characters

$$R_n^{(\alpha_1,\beta_1)}(x) \cdot R_m^{(\alpha_2,\beta_2)}(y) \qquad (-1 < x, y < 1,\ n, m \in \mathbb{N}_0).$$

$J(\boldsymbol{\alpha}, \boldsymbol{\beta})$ is a hypergroup on the square $I \times I$ with $e = (1,1)$.

3.4. Example. Symmetrized product hypergroups

Let $(\alpha, \beta) \in E_J$. Then $J(\alpha, \beta) \otimes J(\alpha, \beta)$ can be symmetrized (c.f. [DR74]) with respect to the reflection $\tau(x, y) = (y, x)$. The resulting hypergroup consists of measures on the triangle with vertices $(1,1)$, $(1,-1)$, and $(-1,-1)$. The polynomials

$$\frac{1}{2}\left[R_n^{(\alpha,\beta)}(x) R_m^{(\alpha,\beta)}(y) + R_m^{(\alpha,\beta)}(x) R_n^{(\alpha,\beta)}(y)\right]$$

are characters, but this family is not algebraically complete since instead of the required two polynomials of degree one, there is only one. An algebraically complete family is obtained by the change of variables $s = (x+y)/2$, $t = xy$. The triangle is transformed to $H = \{(s,t) : 2|s| - 1 \le t \le s^2\}$. The result is a Hermitian continuous 2-variable polynomial hypergroup with $e = (1,1)$.

3.5. Example. Disk polynomial hypergroups

Let $\gamma \ge 0$, $z = re^{i\theta}$, and $D = \{z : |z| \le 1\}$. The polynomials of degree $m + n$

$$R_{n,m}^{\gamma}(z) = r^{|n-m|} e^{i(n-m)\theta} R_{n \wedge m}^{(\gamma, |n-m|)}(2r^2 - 1),$$

(where $n \wedge m$ is the minimum of n and m) are orthogonal on D with respect to the measure $(1 - r^2)^{\gamma} r\, dr\, d\theta$. Then

$$\{R_{n,m}^{\gamma}(z) : n,\ m \in \mathbb{N}_0\}$$

is the character set of a hypergroup on D which we denote $D(\gamma)$ (see [Koo72, AT74, Kan76]). This is an example of a continuous 2-variable polynomial hypergroup that is non-Hermitian since $z^{\vee} = \bar{z}$; $e = 1$.

3.6. Example. Hypergroups on a parabolic biangle

Let $\alpha \geq \beta + 1/2 \geq 0$, and let

$$R_{n,k}^{\alpha,\beta}(x,y) = R_{n-k}^{(\alpha,\beta+k+1/2)}(2x-1) \cdot x^{k/2} R_k^{(\beta,\beta)}(x^{-1/2}y) .$$

Then there is a Hermitian continuous 2-variable continuous polynomial hypergroup $(H,*)$ on the parabolic region

$$H = \{(x,y) : y^2 \leq x \leq 1\}$$

with characters $R_{n,k}^{\alpha,\beta}$ with $e = (1,1)$ [KS95].

3.7. Example. Hypergroups on a triangle

Let $\alpha \geq \beta + \gamma + 1,\ \beta \geq \gamma \geq -1/2$, and let

$$R_{n,k}^{\alpha,\beta,\gamma}(x,y) = R_{n-k}^{(\alpha,\beta+\gamma+2k+1)}(2x-1) \cdot x^k R_k^{(\beta,\gamma)}(x^{-1}y-1) .$$

Then there is a a Hermitian continuous 2-variable polynomial hypergroup $(H,*)$ on the triangular region

$$H = \{(x,y) : y \leq x \leq 1\}$$

with characters $R_{n,k}^{\alpha,\beta,\gamma}$ with $e = (1,1)$ [KS95].

3.8. Example. Hypergroups on a simplex

The previous example can be generalized from the triangle to a k-dimensional simplex

$$H_k = \{(x_1, x_2, \ldots, x_k) : 0 \leq x_k \leq \cdots \leq x_2 \leq x_1 \leq 1\}.$$

These are k-variable continuous polynomial hypergroups which are not products of lower dimensional hypergroups. The polynomials are defined recursively for $k > 3$, $n_1 \geq n_2 \geq \ldots n_k$ by

$$\begin{aligned} &R_{n_1,n_2,\ldots,n_k}^{\alpha_1,\alpha_2,\ldots,\alpha_{k+1}}(x_1,x_2,\ldots,x_k) \\ &\quad = R_{n_1-n_2}^{\alpha_1,\alpha_2+\cdots+\alpha_{k+1}+2n_2+k-1}(2x_1-1) \cdot x_1^{n_2} R_{n_2,\ldots,n_k}^{\alpha_2,\ldots,\alpha_{k+1}}(\frac{x_2}{x_1},\ldots,\frac{x_k}{x_1}), \end{aligned}$$

and for $k = 3$ by

$$R_{n_1,n_2}^{\alpha_1,\alpha_2,\alpha_3}(x_1,x_2) = R_{n_1-n_2}^{(\alpha_1,\alpha_2+\alpha_3+2n_2+1)} x_1^{n_2} R_{n_2}^{(\alpha_2,\alpha_3)}(2\frac{x_2}{x_1}-1).$$

These are the characters of a Hermitian continuous k-variable polynomial hypergroup with $e = (1, 1, \ldots, 1)$ provided

$$\begin{aligned}
\alpha_1 &\geq \alpha_2 + \cdots + \alpha_{k+1} + k - 1,\\
\alpha_2 &\geq \alpha_3 + \cdots + \alpha_{k+1} + k - 2,\\
&\vdots\\
\alpha_{k-1} &\geq \alpha_k + \alpha_{k+1} + 1,\\
\alpha_k &\geq \alpha_{k+1},\\
\alpha_{k+1} &\geq -\tfrac{1}{2}.
\end{aligned}$$

See [KS95].

3.9. Classification of continuous 2-variable polynomial hypergroups

The study of multivariate polynomial hypergroups is in a very early stage. For instance, there is to date some progress in the classification problem of the 2-variable continuous polynomial hypergroups, but there is nothing approaching the complete solution in the one-variable case.

We outline the 2-variable results in [CS95a]. Suppose $(H, *)$ and $(K, \star)$ are continuous 2-variable polynomial hypergroups. If $(H, *)$ and $(K, \star)$ are linearly equivalent, the sets H and K are connected by an affine transformation, so any attempt to classify hypergroups by means of the underlying set must first proceed by defining some canonical hypergroups. There are two such. We make the usual identifications of $\mathbb{R}^2$ with $\mathbb{C}$ and $\mathbf{x} = (x, y) = x + iy = z$. $(H, *)$ is a *canonical Hermitian hypergroup* if $(x, y) \mapsto x$ and $(x, y) \mapsto y$ are characters. $(H, *)$ is a *canonical non-Hermitian hypergroup* if $z \mapsto z$ is a character. Canonical hypergroups have the following properties:

Theorem 3.1.

1. *If $(H, *)$ is a canonical Hermitian hypergroup, then*
   - *(a) The identity element of $(H, *)$ is $(1, 1)$.*
   - *(b) $H \subset I^2$.*
   - *(c) The first degree characters of $(H, *)$ are $(x, y) \mapsto x$ and $(x, y) \mapsto y$.*
2. *If $(H, *)$ is a canonical non-Hermitian hypergroup, then*
   - *(a) The identity element of $(H, *)$ is the complex number 1.*

(b) $z^\vee = \bar{z}$ *for every* $z \in H$.

(c) $H \subset D$.

(d) The first degree characters of $(H, *)$ *are* $z \mapsto z$ *and* $z \mapsto \bar{z}$.

The following theorem shows that these two canonical types suffice:

Theorem 3.2.

(i) If $(H, *)$ *is a Hermitian hypergroup that has exactly two distinct non-constant characters that are first degree polynomials, then it is linearly equivalent to a canonical Hermitian hypergroup.*

(ii) If $(H, *)$ *is a non-Hermitian hypergroup that has exactly two distinct non-constant characters that are first degree polynomials, then it is linearly equivalent to a canonical non-Hermitian hypergroup.*

The hypergroups $J(\boldsymbol{\alpha}, \boldsymbol{\beta})$ and $D(\gamma)$ play a special role in our discussion. If $(\boldsymbol{\alpha}, \boldsymbol{\beta}) \in E_J \times E_J$, then $J(\boldsymbol{\alpha}, \boldsymbol{\beta})$ is a Hermitian hypergroup, but it is not canonical. If we define the affine transformation A by

$$A(x_1, x_2) = \left(R_1^{(\alpha_1,\beta_1)}(x_1),\ R_1^{(\alpha_2,\beta_2)}(x_2)\right),$$

then $J_C(\boldsymbol{\alpha}, \boldsymbol{\beta}) = A(J(\boldsymbol{\alpha}, \boldsymbol{\beta}))$ is a canonical Hermitian hypergroup defined on the rectangle $[-\gamma_1, 1] \times [-\gamma_2, 1]$ with $\gamma_\nu = (\beta_\nu + 1)/(\alpha_\nu + 1)$ for $\nu = 1,\ 2$. $D(\gamma)$ is a canonical non-Hermitian hypergroup defined on the disk D. Corollaries 3.5 and 3.6 show that there are no other canonical 2-variable continuous polynomial hypergroups defined on these sets. The next two theorems arrive at the same conclusion with weaker geometric hypotheses.

Theorem 3.3. *Assume* $(H, *)$ *is a compact canonical Hermitian 2-variable continuous polynomial hypergroup, then* $(H, *) = J_C(\boldsymbol{\alpha}, \boldsymbol{\beta})$ *for some* $(\boldsymbol{\alpha}, \boldsymbol{\beta}) \in E_J \times E_J$ *if and only if* $H_1 = H \cap (I \times \{1\})$ *and* $H_2 = H \cap (\{1\} \times I)$ *are infinite sets.*

Remark 3.1. Example 3.4 shows the condition in Theorem 3.3 cannot be weakened, since in that case H_1 and H_2 each contain only one point. Similarly, in Example 3.6, $H_2 = I$, so it is an infinite set, but H_1 contains only a single point.

We need an additional definition in order to state the next result: Let $\mathbf{x} \cdot \mathbf{y}$ denote the usual inner product in $\mathbb{R}^2$. If $E \subset \mathbb{R}^2$ we say that

$\mathbf{x}_0$ *is a 2-dimensional accumulation point of* E if $\mathbf{x}_0$ is an accumulation point of E, and the only $\mathbf{a} \in \mathbb{R}^2$ satisfying

$$\mathbf{a} \cdot (\mathbf{x} - \mathbf{x}_0) = o\left(\|\mathbf{x} - \mathbf{x}_0\|\right) \quad (\mathbf{x} \to \mathbf{x}_0 \quad \text{with} \quad \mathbf{x} \in E)$$

is $\mathbf{a} = 0$. The identity element is a 2-dimensional accumulation point of H in every one of the examples except the symmetrized product hypergroups (§3.4).

Theorem 3.4. *If* $(H, *)$ *is a canonical non-Hermitian 2-variable continuous polynomial hypergroup, then* $(H, *) = D(\gamma)$ *for some* $\gamma \geq 0$ *if and only if the identity element* 1 *is a 2-dimensional accumulation point of* H, *and* $\{z \in H : |z| = 1\}$ *contains at least seven points.*

The following two obvious corollaries describe the canonical 2-variable hypergroups which are maximal regarding the underlying set.

Corollary 3.5. *If* R *is a rectangle with sides parallel to the coordinate axes and* $(R, *)$ *is a canonical Hermitian 2-variable continuous polynomial hypergroup, then* $(R, *) = J_C(\boldsymbol{\alpha}, \boldsymbol{\beta})$ *for some* $(\boldsymbol{\alpha}, \boldsymbol{\beta}) = (\alpha_1, \beta_1; \alpha_2, \beta_2) \in E_J \times E_J$. *In particular, if* $R = [-\gamma_1, 1] \times [-\gamma_2, 1]$, *then* $\beta_\nu = \gamma_\nu(\alpha_\nu + 1) - 1$ *for* $\nu = 1,\ 2$.

Corollary 3.6 *If* $(D, *)$ *is a canonical non-Hermitian 2-variable continuous polynomial hypergroup, then* $(D, *) = D(\gamma)$ *for some* $\gamma > 0$.

Examples 3.4, 3.6 and 3.7 show that Theorem 3.3 is not exhaustive of the Hermitian continuous 2-variable polynomial hypergroups; we do not know at this time of any examples that show that Theorem 3.4 is not exhaustive for the non-Hermitian case.

The proof of Theorems 3.3 and 3.4 rely heavily on the idea that the characters of certain continuous 2-variable polynomial hypergroups satisfy a pair of second-order linear partial differential equations.

Theorem 3.7. *Let* $(H, *)$ *be a 2-variable continuous polynomial hypergroup and assume that* e *is a 2-dimensional accumulation point of* H. *Then there is a pair of second-order linear partial differential operators* L_1 *and* L_2 *such that for every character* ϕ *of* $(H, *)$ *we have*

$$L_1\phi = \phi_x(e)\phi \quad \textit{and} \quad L_2\phi = \phi_y(e)\phi.$$

The derivation of these equations uses the same ideas as in the first lecture. We believe that the corresponding equations in the higher dimensions will provide the key to identify many hypergroups, but Example 3.4 shows that there are hypergroups to which this theorem does not apply.

4. Research questions

We gather here several questions that arise from the material in the lectures:

Question 1. *Find necessary and sufficient conditions that the characters of a hypergroup are eigenfunctions of a Sturm–Liouville problem. The one-dimensional hypergroups that I know of are all, in fact, of Sturm–Liouville type. Do there exist one-dimensional hypergroups which are not of Sturm–Liouville type? Is there a hypergroup $(J,*)$ which is not equivalent to a Jacobi type hypergroup?*

Question 2. *Theorems 2.3 and 2.4 give rise to the following question: To what extent can the hypothesis that there be a polynomial of* every *degree among the characters of $(H,*)$ be weakened? Perhaps this can be done if "linear change of variables" is replaced by "change of variables". For instance, the even ultraspherical polynomials $\{\mathbb{R}_{2n}^{(\alpha)}(x)\}$ are the characters of a hypergroup on $[0,1]$. But by [Sze67, (4.1.5)] $\mathbb{R}_{2n}^{(\alpha)}(x) = R_n^{(\alpha,-\frac{1}{2})}(2x^2-1)$. Is it possible that a family of polynomials other than the Jacobi polynomials disguised by a change of variables can appear as all or some characters of a hypergroup?*

Question 3. *There is still a "mousehole" left in Theorem 2.4. Is there a hypergroup with polynomial characters which is not equivalent to any $J(\alpha,\beta)$? For the answer to be "yes" it would be necessary that H be a bounded non-compact set and that the polynomials not be orthogonal. Haar measure would necessarily be unbounded.*

Question 4. *Suppose a Sturm–Liouville type hypergroup $(H,*)$ on a compact interval H has a dual structure which is also a hypergroup. This is the same as simply requiring that any two eigenfunctions of a Sturm–Liouville problem can be expressed as a finite linear combination of eigenfunctions with non-negative coefficients. Does this imply that $(H,*)$ is equivalent to $J(\alpha,\beta)$ for some $(\alpha,\beta)\in E_J$?*

Question 5. *There are polynomials besides the Jacobi polynomials which have positive product formulas and give rise to Banach algebras of measures which are not hypergroups. This class needs further investigation. Such polynomials include the generalized Chebyshev polynomials [Lai80] and the q-ultraspherical polynomials [AI83].*

Question 6. *Which plane regions support canonical 2-variable continuous polynomial hypergroups? In particular, are there any canonical non-Hermitian 2-variable continuous polynomial hypergroups aside from $D(\gamma)$ and subgroups of the circle?*

Question 7. *Do all 2-variable continuous polynomial hypergroups have characters which are related to Jacobi polynomials? This is probably the case when e is a 2-dimensional accumulation point of H because then the polynomials must satisfy differential equations (Theorem 3.7). We see from the example of symmetrized product hypergroups that the characters are related to Jacobi polynomials even when e is not a 2-dimensional accumulation point of H.*

Question 8. *Given a measure $m \in M(\mathbb{R}^2)$ can there be more than one 2-variable continuous polynomial hypergroup with Haar measure m; that is, is it ever possible to select two different families of polynomials orthogonal with respect to m such that there are hypergroups with each family of polynomials as characters?*

5. Guide to the references

As mentioned in the introduction, I am including a more extensive bibliography than is required for these notes. The listing below keys the bibliographic items by topic. For a much more extensive list of references the reader should consult the book of Bloom and Heyer [BH95] or Litvinov's survey [Lit87].

Reference works

[BH95], [Edw67], [EMOT53], [GR90], [Hey84a], [Hey86], [Hey89], [Hey91], [Jew75], [Rud62], [Rud74],[Sze67], [Wat66]

Hypergroups (and related systems) in general

[BH95], [BK92], [CGS95], [CS92], [DR74], [Dun73], [Geb89], [Geb95], [Jew75], [Lev64], [Lit87], [Ros77], [Ros78], [Ros95], [Sch74], [Sch77], [Sch88], [Spe78], [Zeu89], [Zeu91]

Ultraspherical and Jacobi polynomials

[AF69], [Ask74], [CMS91], [CS77], [CS79], [CS89], [CS90c], [CS95b], [Gas70], [Gas72], [Hir56a], [Hir56b], [Koo72], [Koo74a]

Other one-variable polynomials

[AAA84], [AI83], [AKR86], [Ask70], [Boc29], [DR74], [Fav35], [Koo78], [Lai80], [Las83], [LO91], [Mar94], [Rah86],[Sho36], [Soa91], [Szw92a], [Szw92b], [Voi90]

Other special functions

[CMS93], [FK73], [FK79], [Fle72], [Koo75a], [Mar89], [Per86], [Sch71]

Disk polynomials

[AT74], [BG90], [BG91], [BG92], [GS96], [HK93], [Kan76], [Kan85]

Multivariate polynomials

[CS95a], [KMT91], [Koo74b], [Koo75b], [KS67], [KS95], [Mac90], [Tra91]

Differential equations

[AT79], [Ché72], [CMS92], [CS90a], [CS90b]

Probability on hypergroups

[BG90], [BG92], [GG87], [Hey84a], [Hey86], [Hey84b], [Hey91], [Voi90]

Miscellaneous

[Bru87], [Dun66], [Mic51]

[AAA84] W. Al-Salam, W. R. Allaway, and R. Askey. Sieved ultraspherical polynomials. *Trans. Amer. Math. Soc.*, 284:41–54, 1984.

[AF69] R. Askey and J. Fitch. Integral representations for Jacobi polynomials and some applications. *J. Math. Anal. Appl.*, 26:411–437, 1969.

[AI83] R. Askey and M. E. H. Ismail. A generalization of ultraspherical polynomials. In P. Erdos, editor, *Studies in Pure Mathematics*, pages 55–78, Boston, 1983. Birkhauser.

[AKR86] R. Askey, T. H. Koornwinder, and M. Rahman. An integral of products of ultraspherical functions and a q-extension. *J. London Math. Soc.*, 33:133–148, 1986.

[Ask70] R. Askey. Linearization of the product of orthogonal polynomials. In R. Gunning, editor, *Problems in Analysis*, pages 223–228, Priceton, NJ, 1970. Princeton University Press.

[Ask74] R. Askey. Jacobi polynomials, I. New proofs of Koornwinder's Laplace type integral representation and Bateman's bilinear sum. *SIAM J. Math. Anal.*, 5:119–124, 1974.

[AT74] H. Annabi and K. Trimèche. Convolution généralisée sur le disque unité. *C. R. Acad. Sc. Paris*, 278:21–24, 1974.

[AT79] A. Achour and K. Trimèche. Opérateurs de translation généralisée associés á un opérateur différentiel singulier sur un intervalle borné. *C. R. Acad. Sci. Paris*, 288:399–402, 1979.

[BG90] M. Bouhaik and L. Gallardo. Une loi des grandes nombres et un théorème limite central pour les chaînes de Markov sur $\mathbf{N}^2$ associés aux polynômes discaux. *C. R. Acad. Sci. Paris*, 310:739–744, 1990.

[BG91] M. Bouhaik and L. Gallardo. A Mehler-Heine formula for disk polynomials. *Indag. Math.*, 1:9–18, 1991.

[BG92] M. Bouhaik and L. Gallardo. Un théorm̀e limite central dans hypergroupe bidimensionnel. *Ann. Inst. H. Poincaré*, 28(1):47–61, 1992.

[BH95] W. R. Bloom and H. Heyer. *Harmonic analysis of probability measures on hypergroups*, volume 20 of *de Gruyter Studies in Mathematics*. de Gruyter, Berlin, New York, 1995.

[BK92] Y. M. Berezanskii and A. A. Kalyuzhnyi. *Harmonic Analysis in Hypercomplex Systems*. Academia Nauk Ukranii, Institut Matematekii, Kiev Nauko Dumka, Kiev, 1992.

[Boc29] S. Bochner. Über Sturm-Liouvillische Polynomesysteme. *Math. Z.*, 29:730–736, 1929.

[Bru87] R. G. M. Brummelhuis. An F. and M. Riesz theorem for bounded symmetric domains. *Ann. Inst. Fourier (Grenoble)*, 37:139–150, 1987.

[CGS95] W. C. Connett, O. Gebuhrer, and A. L. Schwartz, editors. *Applications of hypergroups and related measure algebras*, Providence, R. I., 1995. American Mathematical Society. Contemporary Mathematics, **183**.

[Ché72] H. Chébli. Sur la positivité des opérateurs de "translation généralisée" associés à un opérateur de Sturm-Liouville sur $]0, \infty[$. *C. R. Acad. Sci. Paris*, 275:601–604, 1972.

[CMS91] W. C. Connett, C. Markett, and A. L. Schwartz. Jacobi polynomials and related hypergroup structures. In H. Heyer, editor, *Probability Measures on Groups X, Proceedings Oberwolfach 1990*, pages 45–81, New York, 1991. Plenum.

[CMS92] W. C. Connett, C. Markett, and A. L. Schwartz. Convolution and hypergroup structures associated with a class of Sturm-Liouville systems. *Trans. Amer. Math. Soc.*, 332:365–390, 1992.

[CMS93] W. C. Connett, C. Markett, and A. L. Schwartz. Product formulas and convolutions for angular and radial spheroidal wave functions. *Trans. Amer. Math. Soc.*, 338:695–710, 1993.

[CS77] W. C. Connett and A. L. Schwartz. The theory of ultraspherical multipliers. *Mem. Amer. Math. Soc.*, 183:1–92, 1977.

[CS79] W. C. Connett and A. L. Schwartz. The Littlewood-Paley theory for Jacobi expansions. *Trans. Amer. Math. Soc.*, 251:219–234, 1979.

[CS89] W. C. Connett and A. L. Schwartz. A Hardy-Littlewood maximal inequality for Jacobi type hypergroups. *Proc. Amer. Math. Soc.*, 107:137–143, 1989.

[CS90a] W. C. Connett and A. L. Schwartz. Analysis of a class of probability preserving measure algebras on a compact interval. *Trans. Amer. Math. Soc.*, 320:371–393, 1990.

[CS90b] W. C. Connett and A. L. Schwartz. Positive product formulas and hypergroups associated with singular Sturm-Liouville problems on a compact interval. *Colloq. Math.*, LX/LXI:525–535, 1990.

[CS90c] W. C. Connett and A. L. Schwartz. Product formulas, hypergroups, and Jacobi polynomials. *Bull. Amer. Math. Soc.*, 22:91–96, 1990.

[CS92] W. C. Connett and A. L. Schwartz. Fourier analysis off groups. In A. Nagel and L. Stout, editors, *The Madison symposium on complex analysis*, pages 169–176, Providence, R. I., 1992. American Mathematical Society. Contemporary Mathematics, **137**.

[CS95a] W. C. Connett and A. L. Schwartz. Continuous 2-variable polynomial hypergroups. In O. Gebuhrer W. C. Connett and A. L. Schwartz, editors, *Applications of hypergroups and related measure algebras*, pages 89–109, Providence, R. I., 1995. American Mathematical Society. Contemporary Mathematics, **183**.

[CS95b] W. C. Connett and A. L. Schwartz. Subsets of R which support hypergroups with polynomial characters. to appear in *J. Comput. Appl. Math.*, 1995.

[DR74] C. F. Dunkl and D. E. Ramirez. Krawtchouk polynomials and the symmetrization of hypergroups. *SIAM J. Math. Anal.*, 5:351–366, 1974.

[Dun66] C. F. Dunkl. Operators and harmonic analysis on the sphere. *Trans. Amer. Math. Soc.*, 125:250–263, 1966.

[Dun73] C. F. Dunkl. The measure algebra of a locally compact hypergroup. *Trans. Amer. Math. Soc.*, 179:331–348, 1973.

[Edw67] R. E. Edwards. *Fourier series*, volume I, II of *New York*. Holt, Rinehart and Winston, Inc., 1967.

[EMOT53] A. Erdelyi, W. Magnus, F. Oberhettinger, and F. G. Tricomi. *Higher Transcendental Functions*, volume I. McGraw-Hill Book Company, New York, 1953.

[Fav35] J. Favard. Sur les polynomes de Tchebicheff. *C. R. Acad. Sci. Paris*, 200:2052–2053, 1935. Sér A–B.

[FK73] M. Flensted-Jensen and T. Koornwinder. The convolution structure for Jacobi function expansions. *Ark. Mat.*, 11:245–262, 1973.

[FK79] M. Flensted-Jensen and T. H. Koornwinder. Positive definite spherical functions on a non-compact rank one symmetric space. In P. Eymard, J. Faraut, G. Schiffman, and R. Takahashi, editors, *Analyse harmonique sur les groupes de Lie, II*, pages 249–282. Springer, 1979. Lecture Notes in Math., 739.

[Fle72] M. Flensted-Jensen. Paley-Wiener type theorems for a differential operator connected with symmetric spaces. *Ark. Mat.*, 10:143–162, 1972.

[Gas70] G. Gasper. Linearization of the product of Jacobi polynomials. II. *Canad. J. Math.*, 32:582–593, 1970.

[Gas72] G. Gasper. Banach algebras for Jacobi series and positivity of a kernel. *Ann. of Math.*, 95:261–280, 1972.

[Geb89] O. Gebuhrer. *Analyse harmonique sur les espaces de Gel'fand-Levitan et applications àla theorie des semi-groupes de convolution.* PhD thesis, Universite Louis Pasteur, Strasbourg, France, 1989.

[Geb95] M.-O. Gebuhrer. Bounded measure algebras: A fixed point approach. In O. Gebuhrer W. C. Connett and A. L. Schwartz, editors, *Applications of hypergroups and related measure algebras*, pages 171–190, Providence, R. I., 1995. American Mathematical Society. Contemporary Mathematics, **183**.

[GG87] L. Gallardo and O. Gebuhrer. Marches aléatoires et hypergroupes. *Exposition. Math.*, 5:41–73, 1987.

[GR90] G. Gasper and M. Rahman. *Basic Hypergeometric Series*, volume 35 of *Encyclopedia of Mathematics and its Applications.* Cambridge University Press, Cambridge, 1990.

[GS96] O. Gebuhrer and A. L. Schwartz. Sidon sets and Riesz sets for some measure algebras on the disk. to appearin *Colloq. Math.*, 1996.

[Hey84a] H. Heyer, editor. *Probability Measures on Groups VII, Proceedings Oberwolfach 1983*, volume 1064 of *Lecture Notes in Math.* Springer, Berlin, 1984.

[Hey84b] H. Heyer. Probability theory on hypergroups: a survey. In H. Heyer, editor, *Probability Measures on Groups VII, Proceedings Oberwolfach 1983*, pages 481–550, Berlin, 1984. Springer. Lecture Notes in Math., vol. 1064.

[Hey86] H. Heyer, editor. *Probability Measures on Groups VIII, Proceedings Oberwolfach 1985*, volume 1210 of *Lecture Notes in Math.* Springer, Berlin, 1986.

[Hey89] H. Heyer, editor. *Probability Measures on Groups IX, Proceedings Oberwolfach 1988*, volume 1379 of *Lecture Notes in Math.* Springer, Berlin, 1989.

[Hey91] H. Heyer, editor. *Probability Measures on Groups X, Proceedings Oberwolfach 1990.* Plenum, New York, 1991.

[Hir56a] I. I. Hirschman, Jr. Harmonic analysis and the ultraspherical polynomials. In *Symposium of the Conference on Harmonic Analysis*, Cornell, 1956.

[Hir56b] I. I. Hirschman, Jr. Sur les polynomes ultraspheriques. *C. R. Acad. Sci. Paris*, 242:2212–2214, 1956.

[HK93] H. Heyer and S. Koshi. Harmonic analysis on the disk hypergroup. Mathematical Seminar Notes, Tokyo Metropolitan University, 1993.

[Jew75] R. I. Jewett. Spaces with an abstract convolution of measures. *Adv. in Math.*, 18:1–101, 1975.

[Kan76] Y. Kanjin. A convolution measure algebra on the unit disc. *Tôhoku Math. J. (2)*, 28:105–115, 1976.

[Kan85] Y. Kanjin. Banach algebra related to disk polynomials. *Tôhoku Math. J. (2)*, 37:395–404, 1985.

[Kin63] J. F. C. Kingman. Random walks with spherical symmetry. *Acta Math.*, 109:11–53, 1963.

[KMT91] E. G. Kalnins, W. Miller, Jr., and M. V. Tratnik. Families of orthogonal and biorthogonal polynomials on the N-sphere. *SIAM J. Math. Anal.*, 22:272–294, 1991.

[Koo72] T. H. Koornwinder. The addition formula for Jacobi polynomials, II, the Laplace type integral representation and the product formula. Technical Report TW 133/72, Mathematisch Centrum, Amsterdam, 1972.

[Koo74a] T. Koornwinder. Jacobi polynomials II. An analytic proof of the product formula. *SIAM J. Math. Anal.*, 5:125–137, 1974.

[Koo74b] T. H. Koornwinder. Orthogonal polynomials in two variables which are eigenfunctions of two algebraically independent partial differential operators. I, II, III, IV. *Indag. Math.*, 36:48–58, 59–66, 357–369, 370–381, 1974.

[Koo75a] T. Koornwinder. A new proof of a Paley-Wiener type theorem for the Jacobi transform. *Ark. Mat.*, 13:145–159, 1975.

[Koo75b] T. H. Koornwinder. Two-variable analogues of the classical orthogonal polynomials. In Richard A. Askey, editor, *Theory and Applications of Special Functions*, pages 435–495, New York, 1975. Academic Press, Inc.

[Koo78] T. H. Koornwinder. Positivity proofs for linearization and connection coefficients of orthogonal polynomials satisfying an addition formula. *J. London Math. Soc.*, (2) 18:101–114, 1978.

[KS67] H. L. Krall and I. M. Sheffer. Orthogonal polynomials in two variables. *Ann. Mat. Pura Appl.*, 76:325–376, 1967.

[KS95] T. H. Koornwinder and A. L. Schwartz. Product formulas and associated hypergroups for orthogonal polynomials on the simplex and on a parabolic biangle. preprint, 1995.

[Lai80] T. P. Laine. The product formula and convolution structure for the generalized Chebyshev polynomials. *SIAM J. Math. Anal.*, 11:133–146, 1980.

[Las83] R. Lasser. Bochner theorems for hypergroups and their applications to orthogonal polynomial expansions. *J. Approx. Theory*, 37:311–325, 1983.

[Lev64] B. M. Levitan. *Generalized Translation Operators.* Israel Program for Scientific Translations, Jerusalem, 1964.

[Lit87] G. L. Litvinov. Hypergroups and hypergroup algebras. *J. Soviet Math.*, 38:1734–1761, 1987.

[LO91] R. Lasser and J. Obermaier. On Fejér means with respect to orthogonal polynomials: a hypergroup-theoretic approach. *Progress in Approximation Theory*, pages 551–565, 1991.

[Mac90] I. G. Macdonald. Orthogonal polynomials associated with root systems. In P. Nevai, editor, *Orthogonal polynomials: theory and practice: (proceedings of the NATO Advanced Study Institute on "Orthogonal Polynomials and Their Applications," the Ohio State University, Columbus, Ohio, U.S.A., May 22-June 3, 1989)*, pages 311–318. Kluwer Academic Publishers, 1990.

[Mar89] C. Markett. Product formulas and convolution structure for Fourier-Bessel series. *Constr. Approx.*, 5:383–404, 1989.

[Mar94] C. Markett. Linearization of the product of symmetrical orthogonal polynomials. *Constr. Approx.*, 10:317–338, 1994.

[Mic51] E. Michael. Topologies on spaces of subsets. *Trans. Amer. Math. Soc.*, 71:152–182, 1951.

[Per86] M. Perlstadt. Polynomial analogs of prolate spheroidal wave functions and uncertainty. *SIAM J. Math. Anal.*, 17:242–248, 1986.

[Rah86] M. Rahman. A product formula for the continuous q-Jacobi polynomials. *J. Math. Anal. Appl.*, 118:309–322, 1986.

[Ros77] K. A. Ross. Hypergroups and centers of measure algebras. *Ist. Naz. Alta Mat. (Symposia Math.)*, 22:189–203, 1977.

[Ros78] K. A. Ross. Centers of hypergroups. *Trans. Amer. Math. Soc.*, 243:251–269, 1978.

[Ros95] K. A. Ross. Signed hypergroups – a survey. In O. Gebuhrer W. C. Connett and A. L. Schwartz, editors, *Applications of hypergroups and related measure algebras*, pages 319–329, Providence, R. I., 1995. American Mathematical Society. Contemporary Mathematics, **183**.

[Rud62] W. Rudin. *Fourier Analysis on Groups.* Interscience Publishers, 1962.

[Rud74] W. Rudin. *Real and complex analysis.* McGraw-Hill Book Company, New York, second edition, 1974.

[Sch71] A. L. Schwartz. The structure of the algebra of Hankel and Hankel-Stieltjes transforms. *Canad. J. Math.*, 23:236–246, 1971.

[Sch74] A. L. Schwartz. Generalized convolutions and positive definite functions associated with general orthogonal series. *Pacific J. Math.*, 55:565–582, 1974.

[Sch77] A. L. Schwartz. ℓ^1-convolution algebras: representation and factorization. *Z. Wahrsch. Verw. Gebiete*, 41:161–176, 1977.

[Sch88] A. L. Schwartz. Classification of one-dimensional hypergroups. *Proc. Amer. Math. Soc.*, 103:1073–1081, 1988.

[Sho36] J. Shohat. The relation of the classical orthogonal polynomials to the polynomials of Appell. *Amer. J. Math.*, 58:453–464, 1936.

[Soa91] P. Soardi. Bernstein polynomials and random walks on hypergroups. preprint, 1991.

[Spe78] R. Spector. Mesures invariantes sur les hypergroupes. *Trans. Amer. Math. Soc.*, 239:147–165, 1978.

[Sze67] G. Szegö. *Orthogonal Polynomials*, volume 23 of *Colloquium Publications*. Amer. Math. Soc., Providence, RI, second edition, 1967.

[Szw92a] R. Szwarc. Orthogonal polynomials and a discrete boundary value problem I. *SIAM J. Math. Anal.*, 23:959–964, 1992.

[Szw92b] R. Szwarc. Orthogonal polynomials and a discrete boundary value problem II. *SIAM J. Math. Anal.*, 23:965–969, 1992.

[Tra91] M. V. Tratnik. Some multivariable orthogonal polynomials of the Askey tableau-continuous families. *J. Math. Phys.*, 32:2065–2073, 1991.

[Voi90] M. Voit. Central limit theorems for a class of polynomial hypergroups. *Adv. in Appl. Probab.*, 22:68–87, 1990.

[Wat66] G. Watson. *A Treatise on the Theory of Bessel Functions.* Cambridge University Press, 1966.

[Zeu89] H. Zeuner. One-dimensional hypergroups. *Adv. in Math.*, pages 1–18, 1989.

[Zeu91] H. Zeuner. Duality of commutative hypergroups. In H. Heyer, editor, *Probability Measures on Groups X, Proceedings Oberwolfach 1990*, New York, 1991. Plenum.

Email: schwartz@arch.umsl.edu; Department of Mathematics and Computer Science, University of Missouri–St. Louis, St. Louis, MO 63121

Harmonic Analysis and Functional Equations

Henrik Stetkær

1. Introduction

Functional equations occur in many parts of mathematics, also in harmonic analysis. As an example we mention that the complex exponential function $\gamma : x \to \exp(\xi x)$ for any $\xi \in \mathbf{R}$ is a solution of Cauchy's functional equation

$$\gamma(x+y) = \gamma(x)\gamma(y), \qquad x, y \in \mathbf{R} \tag{1}$$

More generally, the so-called spherical functions play the role in harmonic analysis on homogeneous spaces that the trigonometric polynomials do in Fourier analysis. Also, the spherical functions can be characterized as solutions of certain functional equations. We will in this lecture study a special type of functional equations and point out some of its relations to (1) spherical functions, (2) mean value properties, and (3) addition formulas.

The general set up. Let G be an abelian topological group, and let K be a compact topological group acting as a topological transformation group on G. We write $k \cdot x$ for the action of $k \in K$ on $x \in G$. We assume that K acts by automorphisms of G, so that the map $g \to k \cdot g$ of G into G is an automorphism for each fixed $k \in K$. The action of K on G gives rise to an action of K on the functions on G: If f is a function on G and $k \in K$ we let $(k \cdot f)(x) := f(k^{-1} \cdot x), x \in G$. Finally we let dk denote the normalized Haar measure on the compact group K.

As an example we mention that the integral over $\mathbf{Z}_2 = \{\pm 1\}$ is $\int_{\mathbf{Z}_2} h(k)dk = (h(+1) + h(-1))/2$. In this example $(-1) \cdot x = \sigma x$ for all $x \in G$, where $\sigma : G \to G$ is a continuous automorphism of G such that $\sigma^2 = I$. The classical instance is $\sigma = -I$.

We shall examine certain functional equations of the general form

$$\int_K f(x + k \cdot y)dk = \sum_{l=1}^{N} g_l(x)h_l(y), \qquad x, y \in G, \tag{2}$$

where the functions $f, g_1, \ldots, g_N, h_1, \ldots, h_N \in C(G)$ are to be determined.

For later use we note the following result:

Theorem A. *Let $g_1, \ldots, g_N, h_1, \ldots, h_N \in C(G)$. If the equation (2) has a solution for $f \in C(G)$ then*

$$\sum_{l=1}^{N} \int_K g_l(x + k \cdot y)dk\, h_l(z)$$
$$= \sum_{l=1}^{N} g_l(x) \int_K h_l(y + k \cdot z)dk \qquad \text{for all } x, y, z \in C. \tag{3}$$

Proof. For any $f \in C(G)$ and $x, y, z \in G$ we consider the iterated integral

$$\int_K \int_K f(x + k \cdot y + k' \cdot z)dk'dk = \int_K \int_K f(x + k \cdot [y + (k^{-1}k') \cdot z])dk'dk.$$

When we on the right hand side first introduce $k_1 = k^{-1}k'$ as variable instead of k' and then change order of integration we get the identity

$$\int_K \int_K f(x + k \cdot y + k' \cdot z)dk'dk = \int_K \int_K f(x + k \cdot [y + k_1 \cdot z])dkdk_1.$$

Applying the formula (2) to this identity we obtain (3). ∎

Theorems 3, 6 and 7 have only appeared in preprint form, and Theorem 5 is new.

2. Spherical functions

An example of the functional equation (2) is d'Alembert's functional equation

$$\int_K \phi(x + k \cdot y)dk = \phi(x)\phi(y), \qquad x, y \in G, \tag{4}$$

where $\phi \in C(G)$ is the unknown. For $K = \{I\}$ it reduces to Cauchy's functional equation (1). A particular case for $K = \mathbf{Z}_2$ is the functional equation

$$\frac{\phi(x + y) + \phi(x - y)}{2} = \phi(x)\phi(y), \qquad x, y \in \mathbf{R}, \tag{5}$$

that d'Alembert studied in connection with his investigations of the wave equation [4]. His functional equation (5) is also called the cosine equation, because its non-zero solutions are the functions of the form $\phi(x) = \cos(\lambda x), \lambda \in \mathbf{C}$.

Another interesting instance is the case of $K = SO(2)$ acting in the natural way on $G = \mathbf{R}^2 = \mathbf{C}$. Here d'Alembert's functional equation takes the form

$$\frac{1}{2\pi}\int_0^{2\pi} \phi\left(x + e^{i\theta}y\right) d\theta = \phi(x)\phi(y), \qquad x, y \in \mathbf{C}.$$

It is known that the non-zero (continuous) solutions are described by certain complex Bessel functions (See Proposition IV.2.7 of [11]).

Definition. A K-spherical function on G is a solution $\phi \in C(G), \phi \neq 0$, of d'Alembert's functional equation (4).

In the standard terminology our K-special functions are the spherical functions on the semidirect product $G \times_s K$ with respect to the subgroup K. Note that $\phi = 1$ is a K-spherical function.

Proposition 1. *Any function of the form $\int_K k \cdot \gamma \, dk$, where $\gamma : G \to \mathbf{C}^*$ is a continuous homomorphism, is K-spherical.*

If K is finite then a continuous homomorphism $\gamma_0 : G \to \mathbf{C}^*$ for which $\int_K k \cdot \gamma_0 \, dk = \int_K k \cdot \gamma \, dk$ has the form $\gamma_0 = k_0 \cdot \gamma$ for some $k_0 \in K$.

Proof. Let $\phi := \int_K k \cdot \gamma dk$. Noting that $k \cdot \gamma$ is a homomorphism because γ is, we get by a change of order of integration that

$$\begin{aligned}
\int_K \phi(x + k_1 \cdot y) dk_1 &= \int_K \int_K (k \cdot \gamma)(x + k_1 \cdot y) dk \, dk_1 \\
&= \int_K \int_K (k \cdot \gamma)(x)(k \cdot \gamma)(k_1 \cdot y) dk \, dk_1 \\
&= \int_K (k \cdot \gamma)(x) \left\{ \int_K (k \cdot \gamma)(k_1 \cdot y) dk_1 \right\} dk \\
&= \int_K (k \cdot \gamma)(x) \left\{ \int_K \gamma((k^{-1}k_1) \cdot y) dk_1 \right\} dk.
\end{aligned}$$

The translation invariance of the Haar measure reduces this to

$$\begin{aligned}
\int_K \phi(x + k_1 \cdot y) dk_1 &= \int_K (k \cdot \gamma)(x) \left\{ \int_K \gamma(k_1 \cdot y) dk_1 \right\} dk \\
&= \phi(x) \int_K \gamma(k_1 \cdot y) dk_1 \, .
\end{aligned}$$

By the invariance of the Haar measure under involution we get

$$\int_K \phi(x + k_1 \cdot y)dk_1 = \phi(x) \int_K (k_1^{-1} \cdot \gamma)(y)dk_1$$
$$= \phi(x) \int_K (k_1 \cdot \gamma)(y)dk_1 = \phi(x)\phi(y).$$

The last statement of the proposition is a consequence of the fact that the set of homomorphisms of G into $\mathbf{C}^*$ is linearly independent in the vector space of all complex valued functions on G. (See for example Lemma 29.41 of [12] for a proof.) ■

The following result is derived in [17] by classical harmonic analysis.

Theorem 1. *If ϕ is a K-spherical function on $G = \mathbf{R}^m \times \mathbf{T}^n$ then there exists a continuous homomorphism $\gamma : G \to \mathbf{C}^* = \mathbf{C}\backslash\{0\}$ such that $\phi = \int_K k \cdot \gamma \, dk$.*

For general G very little is known about the solutions of d'Alembert's functional equation (4). However, for $K = \mathbf{Z}_2$ its solution set was found by Kannappan [13] for $\sigma = -I$, and by Baker [6] for any automorphism σ of order 2.

Theorem 2. *Let $\sigma : G \to G$ be a continuous automorphism for G such that $\sigma^2 = I$. Any continuous solution $\phi \neq 0$ of d'Alembert's functional equation*

$$\frac{\phi(x + y) + \phi(x + \sigma y)}{2} = \phi(x)\phi(y), \qquad x, y \in G,$$

may be written in the form $\phi = (\gamma + \gamma \circ \sigma)/2$, where $\gamma : G \to \mathbf{C}^*$ is a continuous homomorphism.

γ is essentially unique: If $\gamma_1 : G \to \mathbf{C}^*$ is a homomorphism and $(\gamma_1 + \gamma_1 \circ \sigma)/2 = (\gamma + \gamma \circ \sigma)/2$ then either $\gamma_1 = \gamma$ or $\gamma_1 = \gamma \circ \sigma$.

Theorem 2 says that the solutions of d'Alembert's functional equation can be expressed by homomorphisms just like the cosine function can be expressed by exponentials via Euler's formulas. Inspired by this, the idea of the proof of Theorem 2 is, for given ϕ, to try to find a corresponding homomorphism γ as a linear combination of ϕ and translates of ϕ. More precisely, it turns out to be a good idea to try

$$\gamma(x) = \phi(x) + k(\phi(x + x_0) - \phi(x + \sigma x_0)), \; x \in G,$$

where $k \in \mathbf{C}$ and $x_0 \in G$ are to be chosen judiciously. For details the reader can consult the proof of Theorem III.1 of [19].

In Theorems 1 and 2 we observe that the basic building blocks of harmonic analysis, i.e. the continuous homomorphisms $\gamma : G \to \mathbf{C}^*$, play a decisive role.

We have not assumed above that the solutions were bounded. The special case of bounded solutions can be handled using classical harmonic analysis. See Chojnacki [7], [8], Badora [5] and Stetkær [15].

d'Alembert's classical functional equation (5) was generalized in 1919 by Wilson [20] to $(f(x+y)+f(x-y))/2 = f(x)\phi(y)$, $x, y \in \mathbf{R}$. We generalize it still further to

$$\int_K f(x+k\cdot y)dk = f(x)\phi(y), \qquad x, y \in G, \tag{6}$$

in which there are 2 unknown functions f and ϕ instead of only one. Actually, this new functional equation (6) also occurs in the theory of spherical functions (see, e.g. Proposition IV.2.4 of [11]). The most general result known for abelian groups is the following which is due to Aczél, Chung and Ng [1] for the case $\sigma = -I$ (see also [16] and [19]).

Theorem 3. *Let $\sigma : G \to G$ be a continuous automorphism of order 2. Let $f, \phi \in C(G)$ satisfy Wilson's functional equation*

$$\frac{f(x+y)+f(x+\sigma y)}{2} = f(x)\phi(y), \qquad x, y \in G. \tag{7}$$

(a) *If $f \neq 0$ then ϕ has the form $\phi = (\gamma + \gamma \circ \sigma)/2$ for some continuous homomorphism $\gamma : G \to \mathbf{C}^*$. We assume from now on that ϕ has this form.*

(b) *If $\gamma \neq \gamma \circ \sigma$ then the set of solutions f of Wilson's functional equation (7) consists of all functions of the form*

$$f = c\frac{\gamma + \gamma \circ \sigma}{2} + c'\frac{\gamma - \gamma \circ \sigma}{2}$$

where c and c' are complex constants.

(c) *If $\gamma = \gamma \circ \sigma$ then the set of solutions f of Wilson's functional equation (7) consists of all functions of the form $f = \gamma(c + a^-)$, where c is a complex constant and $a^- : G \to \mathbf{C}$ is a continuous additive map such that $a^- \circ \sigma = -a^-$.*

Proof. (a) Theorem A gives here that ϕ is a K-spherical function so that Theorem 2 applies.

(b) + (c): It is easy to see that the formulas define solutions, so it is left to prove that any solution f has one of the described forms. Letting $f_\pm := (f \pm f \circ \sigma)/2$ we find that

$$\frac{f_\pm(x+y) + f_\pm(x+\sigma y)}{2} = f_\pm(x)\phi(y), \qquad x, y \in G.$$

Taking $x = 0$ we find $f_+ = f_+(0)\phi$, which is the first term of f under (b) as well as (c). We interchange x and y in

$$\frac{f_-(x+y) + f_-(x+\sigma y)}{2} = f_-(x)\phi(y), \qquad x, y \in G,$$

to obtain

$$\frac{f_-(x+y) - f_-(x+\sigma y)}{2} = f_-(y)\phi(x), \qquad x, y \in G.$$

Adding these two identities yields

$$f_-(x+y) = f_-(x)\phi(y) + \phi(x)f_-(y), \qquad x, y \in G. \tag{8}$$

(c) If $\gamma = \gamma \circ \sigma$ so that $\gamma = \gamma \circ \sigma = \phi$ then we get from (8) that

$$\frac{f_-}{\gamma}(x+y) = \frac{f_-}{\gamma}(x) + \frac{f_-}{\gamma}(y), \qquad x, y \in G,$$

which proves (c).

(b) We will view (8) as a special case of the functional equation (2) by writing it in the form $\int_K f_-(x + k \cdot y)dk = f_-(x)\phi(y) + \phi(x)f_-(y+z)$, where $K = \{I\}$ is the 1 point group. Applying Theorem A to this special case of (2) we get the identity

$$f_-(x+y)\phi(z) + \phi(x+y)f_-(z) = f_-(x)\phi(y+z) + \phi(x)f_-(y+z).$$

We substitute the expressions for $f_-(x+y)$ and $f_-(y+z)$ from (8) into the identity and find after cancellation of some terms that

$$f_-(x)[\phi(y+z) - \phi(y)\phi(z)] = f_-(z)[\phi(x+y) - \phi(x)\phi(y)].$$

Substituting

$$\phi(x+y) - \phi(x)\phi(y) = \frac{\gamma(x) - \gamma(\sigma x)}{2}\frac{\gamma(y) - \gamma(\sigma y)}{2}$$

we get (b). ∎

3. Mean value properties

A special case of Wilson's generalization (6) of d'Alembert's functional equation comes about for $\phi = 1$ where the functional equation reduces to

$$\int_K f(x + k \cdot y)dk = f(x), \qquad x, y \in G. \tag{9}$$

To put things into perspective, one should observe that replacing x by $(\xi + \eta)/2$ and y by $(\xi - \eta)/2$ in the special case

$$\frac{f(x+y) + f(x-y)}{2} = f(x), \qquad x, y \in \mathbf{R},$$

of (9) we get Jensen's classical functional equation

$$\frac{f(\xi) + f(\eta)}{2} = f\left(\frac{\xi + \eta}{2}\right).$$

If the equality sign is replaced by $\geq$ we have Jensen's inequality for convex functions. As a corollary of Theorem 3 we solve the following version of Jensen's functional equation:

Theorem 4. *Let $\sigma : G \to G$ be a continuous automorphism of order 2. The set of solutions $f \in C(G)$ of Jensen's functional equation*

$$\frac{f(x+y) + f(x+\sigma y)}{2} = f(x), \qquad x, y \in G, \tag{10}$$

consists of the functions of the form $f = c + a^-$, where c is a complex constant and $a^- : G \to \mathbf{C}$ is a continuous additive map such that $a^- \circ \sigma = -a^-$.

The functional equation (9) describes a mean value property of f. In the example of $K = SO(2)$, acting in the natural way on $G = \mathbf{R}^2 = \mathbf{C}$, the functional equation says that

$$\int_0^{2\pi} f\left(x + e^{i\theta}y\right) \frac{d\theta}{2\pi} = f(x), \qquad x, y \in \mathbf{C},$$

i.e. the value of f at the center of any circle is the mean of its values on the circle. As is well known this characterizes the harmonic functions.

Functions having the mean value property over other geometric figures than circles, e.g. vertices of regular polygons or rectangles in $\mathbf{R}^2$, have been studied by Aczél et al. [3], Chung et al. [9] and in a more general situation by Stetkær [18].

The functional equation corresponding to regular N-gons is

$$\frac{1}{N}\sum_{n=0}^{N-1} f(x+\omega^n y) = f(x), \qquad x,y \in \mathbf{C},$$

where $\omega = \exp(2\pi i/N)$. The equation is clearly of the form (9) above with $K = \mathbf{Z}_N$. Its set of solutions is span $\{1, z, \dots, z^{N-1}, \bar{z}, \dots, \bar{z}^{N-1}\}$ (see, for example Theorem III.4 of [18]).

The rectangular functional equation which reads

$$\begin{aligned} &f(x_1+y_1, x_2+y_2) + f(x_1+y_1, x_2-y_2) + f(x_1-y_1, x_2+y_2) \\ &\quad + f(x_1-y_1, x_2-y_2) = 4f(x_1,x_2) \qquad \text{for } (x_1,x_2),(y_1,y_2) \in \mathbf{R}^2, \end{aligned} \tag{11}$$

falls into our framework with $K = \mathbf{Z}_2 \times \mathbf{Z}_2$ acting coordinatewise on $\mathbf{R}_2$. Its set of solutions are all functions of the form $f(x_1,x_2) = a_{12}x_1x_2 + a_1x_1 + a_2x_2 + a_0$, where a_{12}, a_1, a_2, a_0 are complex constants [3]. A generalization of it is the functional equation

$$\int_{\mathbf{Z}_2} \cdots \int_{\mathbf{Z}_2} f(x_1 + k_1\cdot y_1, \dots, x_n + k_n \cdot y_n)dk_1 \dots dk_n = f(x_1,\dots,x_n), \tag{12}$$

where $K = \mathbf{Z}_2 \times \cdots \times \mathbf{Z}_2$ acts coordinatewise on $G = G_1 \times \cdots \times G_n$.

Let us for each $i = 1,\dots,n$ write the action of $-1 \in \mathbf{Z}_2 = \{\pm 1\}$ on G_i as $(-1)\cdot x_i = \sigma_i x_i$ so that $\sigma_i : G_i \to G_i$ is a continuous involution. For $\sigma_i = -I, i = 1,\dots,n$, the following theorem is derived by a long induction on n in [9]. Our result generalizes not just the formulas for the solutions of the rectangular functional equations (11) and (12) of [3] and [9] but also the special case of Jensen's functional equation from Theorem 4 above.

Theorem 5. *The set of solutions of (12) consists of the linear span of the functions of the form* $(x_1,\dots,x_n) \to a_{p_1\cdots p_l}(x_{p_1},\dots,x_{p_l})$ *with* $1 \le p_1 < \cdots < p_l \le n, 0 \le l \le n$, *where* $a_{p_1\cdots p_l} : G_{p_1} \times \cdots \times G_{p_l} \to \mathbf{C}$ *ranges over the continuous multi-additive functions satisfying*

$$a_{p_1\cdots p_l}(x_{p_1},\dots,\sigma_{p_j}x_{p_j},\cdots,x_{p_l}) = -a_{p_1\cdots p_l}(x_{p_1},\dots,x_{p_j},\dots,x_{p_l}) \tag{13}$$

for $j = 1,\dots,l$. *If* $l = 0$ *the function shall be interpreted as the constant function 1.*

Proof. It is elementary to verify that any function of the described form is a solution of (12), so it is left to show that any solution f has this form.

Let $K = \mathbf{Z}_2 \times \cdots \times \mathbf{Z}_2$. It is known from the theory of representations of compact groups (see, e.g., Lemma IV.1.9 of [11]) that the decomposition of a function into an even and an odd part has an extension to the present set up: Any $f \in C(G)$ may be written as $f = \sum_{\chi \in \hat{K}} f_\chi$ where $f_\chi = \int_K k \cdot f \overline{\chi(k)}\, dk$ has the property $k \cdot f_\chi = \chi(k) f_\chi, k \in K$. As is easy to see, f_χ satisfies the functional equation (12), so replacing f by f_χ we may from now on assume that f satisfies the invariance condition $k \cdot f = \chi(k) f, k \in K$, for some homomorphism $\chi : \mathbf{Z}_2 \times \cdots \times \mathbf{Z}_2 \to \mathbf{T}$. Any such homomorphism has the form

$$\chi(k_1, k_2, \ldots, k_n) = \chi_1(k_1)\chi_2(k_2) \cdots \chi_n(k_n), \ (k_1, \ldots, k_n) \in K, \qquad (14)$$

where $\chi_1 : \mathbf{Z}_2 \to \mathbf{T}$ is a homomorphism for each $i = 1, 2, \ldots, n$.

It follows from (12) that f as a function of its i'th variable is a solution of Jensen's functional equation (10), and from the invariance property that $k_i \cdot f = \chi_i(k_i) f$. From Theorem 1 we get that f does not depend on its i'th variable if $\chi_i = 1$, and that f is additive and satisfies (13) in its i'th variable if $\chi_i \neq 1$. ∎

4. Addition formulas

The topic that I want to discuss last is the addition formulas of sine and cosine. We will show that these formulas are closely related.

If $K = \{I\}$ and $G = \mathbf{R}$ then the pair $f = \sin$ and $g = \cos$ constitute a solution of the functional equation

$$\int_K f(x + k \cdot y) dk = f(x)g(y) + g(x)f(y), \qquad x, y \in G. \qquad (15)$$

This is just the addition formula for sine. The addition formula for cosine is

$$\int_K g(x + k \cdot y) dk = g(x)g(y) - f(x)f(y), \qquad x, y \in G. \qquad (16)$$

Theorem 6. *If $f, g \in C(G)$ constitute a solution of the functional equation (15) and $f \neq 0$, then there exists a constant $\kappa \in \mathbf{C}$ such that*

$$\int_K g(x + k \cdot y) dk = g(x)g(y) + \kappa f(x)f(y), \qquad x, y \in G.$$

Furthermore both f and g are K-invariant in the sense that $f(k \cdot x) = f(x)$ and $g(k \cdot x) = g(x)$ for all $k \in K$ and $x \in G$.

In other words, g must satisfy the addition formula for cosine. For $K = \{I\}$ the result can be found as Proposition 13.2 of [2].

Proof. We first prove the statement on K-invariance. Let $k_0 \in K$. Replacing y by $k_0 \cdot y$ in (15) we get by the translation invariance of the Haar measure on K that $f(x)g(k_0 \cdot y) + g(x)f(k_0 \cdot y) = f(x)g(y) + g(x)f(y)$. The statement follows if f and g are linearly independent. If they are not then $g = \mu f$ for some $\mu \in \mathbf{C}$ and the functional equation reduces to $\int_K f(x + k \cdot y)dk = 2\mu f(x)f(y)$. Now, $\mu \neq 0$, because $\mu = 0$ implies that $f = 0$ contradicting the assumption. Taking $x_0 \in G$ such that $f(x_0) \neq 0$ we find that

$$f(y) = \frac{1}{2\mu f(x_0)} \int_K f(x_0 + k \cdot y)dk.$$

This formula shows that f is K-invariant. Hence so is $g = \mu f$. Observe for use below that

$$\int_K g(y + k \cdot x)dk = \int_K g(k^{-1} \cdot y + x)dk = \int_K g(x + k \cdot y)dk$$

by the invariance of the Haar measure under the change of variables $k \to k^{-1}$.

Theorem A applied to the functional equation shows that

$$f(x) \int_K g(y + k \cdot z)dk = g(z) \int_K f(y + k \cdot z)dk = [f(x)g(y) + g(x)f(y)]g(z),$$

so that $f(x)[\int_K g(y + k \cdot z)dk - g(y)g(z)] = g(x)f(y)g(z)$. The right hand side is unchanged if x and z change places, hence so is the left hand side, i.e. $f(x)[\int_K g(y + k \cdot z)dk - g(y)g(z)] = f(z)[\int_K g(y + k \cdot z)dk - g(y)g(x)]$. Applying the observation above to the right hand side we find that

$$f(x) \left[\int_K g(y + k \cdot z)dk - g(y)g(z) \right] = f(z) \left[\int_K g(x + k \cdot y)dk - g(x)g(y) \right]. \tag{17}$$

Now, $f(z_0) \neq 0$ for some $z_0 \in G$. Putting $z = z_0$ and dividing through by $f(z_0)$ we see that there is a function $\psi : G \to \mathbf{C}$ such that

$$\int_K g(x + k \cdot y)dk - g(x)g(y) = f(x)\psi(y), \qquad x, y \in G.$$

Introducing this expression into (17) gives that $f(x)f(y)\psi(z) = f(z)f(x)\psi(y)$ from which we see that ψ is a constant multiple of f.

■

Theorem 7. *If $f, g \in C(G)$ constitute a solution of the functional equation (16) then there exists a constant $\alpha \in \mathbf{C}$ such that*

$$\int_K f(x + k \cdot y)dk = f(x)g(y) + g(x)f(y) + \alpha f(x)f(y), \qquad x, y \in G.$$

Furthermore both f and g are K-invariant in the sense that $f(k \cdot x) = f(x)$ and $g(k \cdot x) = g(x)$ for all $k \in K$ and $x \in G$.

Proof. The proof of the K-invariance of f and g goes along the same lines as in Theorem 6, so we skip it.

If f and g satisfy (16), Theorem A tells us that

$$\begin{aligned}\int_K g(x + k \cdot y)dk g(z) - \int_K f(x + k \cdot y)dk f(z) \\ = g(x) \int_K g(y + k \cdot z)dk - f(x) \int_K f(y + k \cdot z)dk\end{aligned}$$

so that by (16)

$$\begin{aligned}[g(x)g(y) - f(x)f(y)]g(z) - \int_K f(x + k \cdot y)dk f(z) \\ = g(x)[g(y)g(z) - f(y)f(z)] - f(x) \int_K f(y + k \cdot z)dk\end{aligned}$$

or

$$f(x) \int_K f(y+k \cdot z)dk - f(x)f(y)g(z) = f(z) \int_K f(x+k \cdot y)dk - g(x)f(y)f(z).$$

Subtract $f(x)g(y)f(z)$ from both sides to get

$$\begin{aligned}f(x) \left[\int_K f(y + k \cdot z)dk - f(y)g(z) - g(y)f(z)\right] \\ = f(z) \left[\int_K f(x + k \cdot y)dk - f(x)g(y) - g(x)f(y)\right]. \qquad (18)\end{aligned}$$

If $f = 0$ the conclusion of the theorem is trivially true. Thus we can assume that there is $z_0 \in G$ at which $f(z_0) \neq 0$. Now let

$$\phi(x, y) := \int_K f(x + k \cdot y)dk - f(x)g(y) - g(x)f(y).$$

Using $z = z_0$ the identity (18) now reads $f(x)\phi(y, z_0) = f(z_0)\phi(x, y), x, y \in G$, so that

$$\phi(x, y) = f(x)\psi(y), \text{ where } \psi(y) := \phi(y, z_0)/f(z_0). \qquad (19)$$

Using $z = z_0$ the identity (18) now reads $f(x)\phi(y, z_0) = f(z_0)\phi(x, y), x, y \in G$, so that

$$\phi(x, y) = f(x)\psi(y), \quad \text{where} \quad \psi(y) := \phi(y, z_0)/f(z_0). \tag{19}$$

Substitute (19) into (18) to get $f(z_0)f(x)\psi(y) = f(x)f(y)\psi(z_0)$ or $f(z_0)\psi(y) = f(y)\psi(z_0)$. Thus $\psi(y) = \alpha f(y)$, and $\phi(x, y) = \alpha f(x)f(y)$, proving the theorem. ■

Much more should be known about the addition formulas (15) and (16) for a general action of a compact group K, but to the best of my knowledge only the cases of $K = \{I\}$ and $K = \mathbf{Z}_2$ have been solved completely (see [10] and [14]). The solutions involve group homomorphisms $\gamma : G \to \mathbf{C}^*$ so harmonic analysis should intervene in the general set up.

I hope in this lecture to have convinced you that there are interesting and important relations between the basic building blocks of harmonic analysis and functional equations, and that the topic is far from being completely explored.

References

[1] Aczél, J., Chung, J. K. and Ng, C. T.: *Symmetric second differences in product form on groups.* Topics in mathematical analysis (pp. 1–22) edited by Th. M. Rassias. Ser. Pure Math., 11, World Scientific Publ. Co., Teaneck, NJ, 1989.

[2] Aczél, J. and Dhombres, J.: *Functional equations in several variables.* Cambridge University Press Cambridge/New York/New Rochelle/ Melbourne/Sydney 1989.

[3] Aczél, J., Haruki, H., McKiernan, M. A. and Sakovič, G. N.: *General and Regular Solutions of Functional Equations Characterizing Harmonic Polynomials.* Aequationes Math. **1** (1968), 37–53.

[4] d'Alembert, J., *Recherches sur la courbe que forme une corde tendue mise en vibration, I-II.* Hist. Acad. Berlin (1747) 214–249.

[5] Badora, R.,*On a joint generalization of Cauchy's and d'Alembert's functional equations.* Aequationes Math. **43** (1992), 72–89.

[6] Baker, J. A., *The stability of the cosine equation.* Proc. Amer. Math. Soc. **80** (1980), 411–416.

[7] Chojnacki, W., *Fonctions cosinus hilbertiennes bornées dans les groupes commutatifs localement compacts.* Compositio Math. **57** (1986), 15–60.

[8] Chojnacki, W., *On some functional equations generalizing Cauchy's and d'Alembert's functional equations.* Colloq. Math. **55** (1988), 169–178.

[9] Chung, J. K. (Zhong Jukang), Ebanks, B. R., Ng, C. T., Sahoo, P. K. and Zeng, W. B., *On generalized rectangular and rhombic functional equations.* Publ. Math. (Debrecen) **47** (1995), 249–270.

[10] Chung, J. K., Kannappan, Pl. and Ng, C. T., *On two trigonometric functional equations.* Mathematics Reports Toyama University **11** (1988), 153–165.

[11] Helgason, S.: *"Groups and Geometric Analysis."* Academic Press, Inc., Orlando - San Diego - San Francisco - New York- London - Toronto - Montreal - Sydney - Tokyo - São Paulo 1984.

[12] Hewitt, E. and Ross, K. A.: *"Abstract Harmonic Analysis II".* Springer-Verlag. Berlin-Heidelberg-New York 1970.

[13] Kannappan, Pl., *The functional equation* $f(xy) + f(xy^{-1}) = 2f(x)f(y)$ *for groups.* Proc. Amer. Math. Soc. **19** (1968), 69–74.

[14] Poulsen, T. and Stetkær, H., *Functional equations on abelian groups with involution.* Preprint Series 1995 No **16**, Matematisk Institut, Aarhus University, Denmark, pp. 1–23.

[15] Stetkær, H., *D'Alembert's equation and spherical functions.* Aequationes Math. **48** (1994), 220–227.

[16] Stetkær, H., *Wilson's Functional Equations on Groups.* Aequationes Math. **49** (1995), 252–275.

[17] Stetkær, H., *Functional Equations and Spherical Functions.* Preprint Series 1994 No **18**, Matematisk Institut, Aarhus University, Denmark, pp. 1–28.

[18] Stetkær, H., *Wilson's functional equation on* **C**. Preprint Series 1995 No **1**, Matematisk Institut, Aarhus University, Denmark. pp. 1–15. Accepted for publication by Aequationes Math.

[19] Stetkær, H., *Functional equations on abelian groups with involution.* Accepted for publication by Aequationes Math.

[20] Wilson, W. H., *On certain related functional equations.* Bull. Amer. Math. Soc. **26** (1919), 300–312.

E-mail: stetkaer@mi.aau.dk; Department of Mathematics, Aarhus University, Ny Munkegade, DK 8000 Aarhus C, Denmark

Actions of Finite Hypergroups and Examples

V.S. Sunder and N.J. Wildberger

Abstract

This paper is an introduction to the theory of actions of finite hypergroups, particularly commutative ones. We present some basic facts concerning actions and then proceed to classify irreducible $*$-actions of hypergroups of order two, the class and character hypergroups of S_3 and of the Golden hypergroup —which arises from the pentagon when viewed as a strongly regular graph.

1. Introduction

This paper will try to illustrate the idea of an *action* of a finite hypergroup $\mathcal{K} = \{c_0, \ldots, c_n\}$ on a finite set X through the investigation of some specific examples. The study of actions of objects related to hypergroups was initiated in [11] in the study of invariants associated to inclusions of II_1 factors. We will adapt the notion studied there to finite hypergroups in the sense of Dunkl [4], Jewett [6] and Spector [9]; see also [2], [7], [8].

An action is a special type of representation; one where the representing matrices are all column stochastic, that is with non-negative real entries and column sums 1. If the action preserves the hypergroup involution *, that is if c_i^* is mapped to the adjoint of c_i for all i, then we say we have a $*$-action. In this case the representing matrices will be doubly stochastic. This is the most natural situation from some points of view, and we will be concentrating on it.

An action can be thought of geometrically as an assignment to each c_i of an affine map of some simplex to itself. A $*$-action is a special type of action for which in particular the centroid of the simplex is left fixed by all c_i.

An action or $*$-action is irreducible if no face of the simplex is left stable by all c_i. This definition has no counterpart in representation theory; thus it is possible for even a commutative hypergroup to have irreducible $*$-actions of dimension larger than one. However one of the fundamental facts, as given by Theorem 5, is that this dimension is bounded by the total weight $w(\mathcal{K})$ of the hypergroup.

After giving the main definitions of hypergroups, actions and $*$-actions we discuss some general families of examples arising from quotient hypergroups and class hypergroups of finite groups. We then prove the main inequality mentioned above for the dimension of an irreducible $*$-action of a commutative hypergroup and discuss some related results. We introduce the notion of a maximal action and state a relation between maximal $*$-actions and association schemes.

The last section is devoted to the classification of irreducible $*$-actions of certain commutative hypergroups. In particular, we consider the general hypergroup with two elements, the class and character hypergroups of the symmetric group S_3, and the so-called Golden hypergroup.

Some of the results stated here are proved in a more general context in [12]. The present paper is intended to be an introduction to this circle of ideas; there are a number of obvious questions which we do not at present know how to answer. Among these are the following.

Problem 1. Given a commutative hypergroup $\mathcal{K}$, does $\mathcal{K}$ admit only a finite number of irreducible $*$-actions?

Problem 2. How can one determine these irreducible $*$-actions? In particular, is there any 'internal' description of them?

Problem 3. Which hypergroups admit a *maximal* $*$-action, that is, one where the dimension is exactly $w(\mathcal{K})$? (A partial answer is given in Theorem 7).

Problem 4. The class hypergroup $\mathcal{K}(G)$ for any finite group always admits a maximal $*$-action. Is this unique? How about if G is simple?

2. Definitions of hypergroups and actions

Definition 1. A *finite hypergroup* is a pair $(\mathcal{K}, A)$ where A is a $*$-algebra with unit c_0 over $\mathbb{C}$ and $\mathcal{K} = \{c_0, c_1, \ldots, c_n\}$ is a subset of A satisfying

(A1) $\mathcal{K}$ is a basis of A

(A2) $\mathcal{K}^* = \mathcal{K}$

(A3) The structure constants $n_{ij}^k \in \mathbb{C}$ defined by

$$c_i c_j = \sum_k n_{ij}^k c_k$$

satisfy the conditions

$$c_i^* = c_j \Leftrightarrow n_{ij}^0 > 0$$
$$c_i^* \neq c_j \Leftrightarrow n_{ij}^0 = 0.$$

(A4) $n_{ij}^k \geq 0$
(A5) $\sum_k n_{ij}^k = 1$

$\mathcal{K}$ is called *Hermitian* if $c_i^* = c_i$ for all i and *commutative* if $c_i c_j = c_j c_i$ for all i, j. Hermitian hypergroups are always commutative. In this paper, we will consider commutative hypergroups primarily.

If we replace axiom $(A4)$ with

$(A4')$ $n_{ij}^k \in \mathbb{R}$

then we say $\mathcal{K}$ is a *signed hypergroup.*

Definition 2. The *weight* of an element $c_i \in \mathcal{K}$ is $w(c_i) = (n_{ji}^0)^{-1}$ where $c_j = c_i^*$. The *total weight* of $\mathcal{K}$ is $w(\mathcal{K}) = \sum_{i=0}^n w(c_i)$.

Note that this definition also holds for signed hypergroups and that $w(c_i) \geq 1$. Clearly c_0 has weight 1; the set of elements of weight 1 forms a group.

Two important examples of commutative hypergroups are associated with any finite group G, the *class hypergroup* $\mathcal{K}(G)$ consisting of probability measures on the conjugacy classes under convolution and the *character hypergroup* $\mathcal{K}(G^\wedge)$ consisting of normalized characters of G under pointwise multiplication. Here normalized means $\chi(e) = 1$.

For a finite set $X = \{x_1, \ldots, x_k\}$ let sX denote the collection of all formal sums of the form $\alpha_1 x_1 + \cdots + \alpha_k x_k$ where $\alpha_i \geq 0$ and $\alpha_1 + \cdots + \alpha_k = 1$. We assume no relations among the x_i; that is two sums $\alpha_1 x_1 + \cdots + \alpha_k x_k$ and $\beta_1 x_1 + \cdots + \beta_k x_k$ are equal if and only if $\alpha_i = \beta_i$, $i = 1, \ldots, k$. The set sX is an abstract simplex with vertices X.

For any $x, y \in sX$ and $\alpha \in [0,1]$, the affine combination $\alpha x + (1-\alpha)y$ is a well-defined element of sX and more generally for any $x_1, \ldots, x_n \in sX$ and $\alpha_1, \ldots, \alpha_n \geq 0$, $\alpha_1 + \cdots + \alpha_n = 1$, the element $\alpha_1 x_1 + \cdots + \alpha_n x_n$ is a well-defined element of sX. Furthermore sX contains the distinguished point

$$c = \frac{1}{k} x_1 + \cdots + \frac{1}{k} x_k$$

which we call the *centroid* of sX.

Definition 3. Let $Aff(X)$ denote the set of all maps $\psi : sX \to sX$ that satisfy

$$\psi(\alpha x + (1-\alpha)y) = \alpha\psi(x) + (1-\alpha)\psi(y) \quad \forall\, x, y \in sX,\ \alpha \in [0,1].$$

Let $Aff(X, c)$ be the set of all $\psi \in Aff(X)$ that in addition satisfy

$$\psi(c) = c\ .$$

To any $\psi \in Aff(X)$ we may associate the $k \times k$ nonnegative matrix $T_\psi = [t_{ij}]$ where

$$\psi(x_j) = \sum_{i=1}^{n} t_{ij} x_j.$$

Clearly the column sums of T_ψ are 1, and if $\psi \in Aff(X,c)$ then in addition the row sums are also 1 . The map $\psi \to T_\psi$ is thus a bijection from $Aff(X)$ to the set of column stochastic $k \times k$ matrices which restricts to a bijection from $Aff(X,c)$ to the set of all doubly stochastic $k \times k$ matrices. The latter, which we denote by Ω_k, is closed under affine combinations, multiplication, and adjoints $T \to T^*$. It follows that we may define ψ^* for any $\psi \in Aff(X,c)$ by

$$T_{\psi^*} = (T_\psi)^*$$

so that $Aff(X,c)$ is also closed under affine combinations, multiplication, and adjoints.

Let $\mathcal{K} \subset \mathbb{A}$ be a finite hypergroup. Then sK may be regarded explicitly as the convex hull of $\mathcal{K}$ in $\mathbb{A}$ and is closed under convex linear combinations, multiplication, and adjoints (given by *). We come now to the main definition.

Definition 4. An *action* of a hypergroup $\mathcal{K} = \{c_0, c_1, \ldots, c_n\}$ on a finite set X is a homomorphism $\pi : sK \to Aff(X)$, that is, an assignment to every $c_i \in \mathcal{K}$ of an element $\pi(c_i) \in Aff(X)$ such that

(i) $\pi(c_0)$ is the identity

(ii) $\pi(c_i)\pi(c_j) = \sum_k n_{ij}^k \pi(c_k)$ where n_{ij}^k are the structure constants of $\mathcal{K}$.

If in addition the image of π is in $Aff(X,c)$ and

(iii) $\pi(c_i^*) = \pi(c_i)^*$ for all $c_i \in \mathcal{K}$
then we say that π is a $*$-*action.*

Definition 5. The *character* of an action $\pi : \mathcal{K} \to Aff(X)$ is the function

$$\chi(c_i) = \text{tr } \pi(c_i).$$

which we also write as the vector $\chi = (\chi(c_0), \ldots, \chi(c_n))$.

Note the useful fact that the character must always have non-negative entries.

Definition 6. An action π of $\mathcal{K}$ on X is *reducible* if there exists a nonempty subset $Y \subset X$, $Y \neq X$ such that $\pi(c_i)(Y) \subseteq sY$ $\forall c_i \in \mathcal{K}$. Otherwise π is *irreducible.*

Definition 7. Two actions π_1 and π_2 of $\mathcal{K}$ on sets X_1 and X_2 respectively are *equivalent* if there exists a bijection $\psi : sX_1 \to sX_2$ such that $\psi \circ \pi_1(c_i) = \pi_2(c_i) \circ \psi$ for all $c_i \in \mathcal{K}$.

Recall that for any hypergroup $\mathcal{K} = \{c_0, c_1, \ldots, c_n\}$ the element

$$e_0 = \frac{1}{w(\mathcal{K})} \sum_{i=0}^{n} w(c_i) c_i$$

of $s\mathcal{K}$, called *Haar-measure*, satisfies $c_i e_0 = e_0$ $\forall i$. The following is a restatement into this context of a result of [11].

Theorem 1. *Suppose π is an action of a hypergroup $\mathcal{K}$ on a finite set X. Then the following are equivalent:*
(i) π is irreducible
(ii) $T_{\pi(e_0)}$ is a strictly positive matrix
(iii) $T_{\pi(e_0)}$ is a rank one projection.

Let J_k denote the $k \times k$ matrix consisting of all 1's and let I_k denote the $k \times k$ identity matrix.

Corollary 2. *If π is an irreducible $*$-action of a hypergroup $\mathcal{K}$ of dimension k then $T_{\pi(e_0)} = \frac{1}{k} J_k$.*

Proof. By Theorem 1 and the fact that π is a $*$-action, $T_{\pi(e_0)}$ is a rank one $k \times k$ doubly stochastic matrix. But there is only one such matrix. ■

3. Some general classes of examples

Example 1. Any hypergroup $\mathcal{K}$ acts on itself by left multiplication. This is a $*$-action if and only if $\mathcal{K}$ is a group.

Example 2. For any hypergroup $\mathcal{K} = \{c_0, c_1, \ldots, c_n\}$ and any subhypergroup $\mathcal{L}$, we may form the quotient hypergroup $\mathcal{K}/\mathcal{L}$. Then $\mathcal{K}$ acts on the set $\mathcal{K}/\mathcal{L}$.

We provide some details. A subhypergroup $\mathcal{L}$ of $\mathcal{K}$ is a subset of $\mathcal{K}$ that contains the identity c_0 and is closed under * and multiplication. The latter condition may be conveniently restated as follows. For $c_i, c_j, c_k \in \mathcal{K}$ let us write $c_k \in c_i c_j$ iff $n_{i,j}^k > 0$. Then $\mathcal{L}$ is closed under multiplication iff for any $l_i, l_j \in \mathcal{L}$, $c_k \in l_i l_j$ implies $c_k \in \mathcal{L}$.

Following [6], for $c_i \in \mathcal{K}$ the set

$$c_i\mathcal{L} = \{c_k | c_k \in c_i l_j \text{ for some } l_j \in \mathcal{L}\}$$

is called a *left coset* of $\mathcal{L}$. Left cosets are identical or disjoint. The reason is as follows. If $c_j \in c_i\mathcal{L}$ then $c_j \in c_i l_k$ for some $l_k \in \mathcal{L}$ so that $c_0 \in c_i l_k c_j^*$. Taking the involution * of both sides, we see that $c_0 \in c_j l_k^* c_i^*$ and thus $c_i \in c_j\mathcal{L}$.

The set of all left cosets we denote by $\mathcal{K}/\mathcal{L}$. This set is a hypergroup in a natural way by defining

$$(c_i\mathcal{L})(c_j\mathcal{L}) = \sum_{k=0}^{n} n_{i,j}^k (c_k\mathcal{L})$$

where the latter sum can be rewritten as an affine combination of distinct cosets by collecting common terms. There is then a natural hypergroup homomorphism $p : \mathcal{K} \to \mathcal{K}/\mathcal{L}$ defined by $p(c_i) = c_i\mathcal{L}$. From Example 1 above we see that $\mathcal{K}/\mathcal{L}$ acts on itself; combining this with p shows that $\mathcal{K}$ also acts on $\mathcal{K}/\mathcal{L}$. More detail on the relation between $\mathcal{K}$ and $\mathcal{K}/\mathcal{L}$ may be found in ([14]).

The actions of $\mathcal{K}$ obtained this way have a particular property which we abstract as follows.

Definition 10. The action π of $\mathcal{K}$ on a set X is *strongly transitive* if there exists an $x \in X$ such that $X = \pi(\mathcal{K})\, x$.

Proposition 3. *Given a hypergroup $\mathcal{K}$ and a subhypergroup $\mathcal{L}$, the action of $\mathcal{K}$ on $\mathcal{K}/\mathcal{L}$ is strongly transitive. Conversely any strongly transitive action of $\mathcal{K}$ arises this way from some subhypergroup $\mathcal{L} \subset \mathcal{K}$.*

Proof. The first statement is obvious; just take x to be the coset $\mathcal{L}$. Suppose then that $\pi : \mathcal{K} \to Aff(X)$ is a strongly transitive action on a set X with distinguished element x as in Definition 10. Let $\mathcal{L} = \{c_i \in \mathcal{K} | \pi(c_i)\, x = x\}$. This is a subhypergroup of $\mathcal{K}$. Consider the affine map $\psi : s(\mathcal{K}/\mathcal{L}) \to sX$ given on the vertices by $\psi(c_i\mathcal{L}) = \pi(c_i)\, x$. This map is well defined and estabishes a bijection between the two sets which commutes with the actions of $\mathcal{K}$. ■

Example 3. Let G be a finite group with conjugacy classes $C_0 = \{e\}, C_1, \ldots, C_n$ and let $\mathcal{K}(G) = \{c_0, c_1, \ldots, c_n\}$ be the class hypergroup; recall that this means that c_i is the element $\frac{1}{|C_i|}\sum_{g\in C_i} g$ in the group algebra and that c_i^* is $\frac{1}{|C_i|}\sum_{g\in C_i} g^{-1}$. Suppose that G acts on a finite set $X = \{x_1, \ldots, x_k\}$. Define $\pi(c_i) \in Aff(X, c)$ by

$$\pi(c_i)(x_j) = \frac{1}{|C_i|}\sum_{g\in C_i} g \cdot x_j.$$

Proposition 4. *π is a $*$-action of $\mathcal{K}(G)$ on X. It is irreducible if and only if the action of G on X is transitive.*

Proof. Clearly $\pi(c_0)$ is the identity.

$$\begin{aligned} \pi(c_i)\pi(c_j)(x_v) &= \frac{1}{|C_i|}\frac{1}{|C_j|}\sum_{h\in C_i}\sum_{g\in C_j} h\cdot(g\cdot x_v) \\ &= \frac{1}{|C_i|}\frac{1}{|C_j|}\left(\sum_{h\in C_i}\sum_{g\in C_j} hg\right)\cdot x_v \\ &= \sum_{k=0}^{n} n_{ij}^k \left(\frac{1}{|C_k|}\sum_{g\in C_k} g\right) x_v \\ &= \sum_{k=0}^{n} n_{ij}^k \pi(c_k)(x_v) \end{aligned}$$

so π is a homomorphism. The entry $\pi(c_i)_{uv}$ (identifying the operator with its matrix with respect to the ordered basis $\{x_1, \ldots, x_n\}$) is the coefficient of x_u in the expression $\frac{1}{|C_i|}\sum_{g\in C_i} g\cdot x_v$. But this is the same as the coefficient of x_v in the sum $\frac{1}{|C_i|}\sum_{g\in C_i} g^{-1}x_u$ which is $\pi(c_i^*)_{vu}$. Thus

$$\pi(c_i^*) = \pi(c_i)^*$$

so π is a $*$-action.

If $Y \subset X$ is a non trivial G orbit then $\pi(c_i)Y \subset sY$ for all i so that if the $*$-action π is irreducible then G acts transitively. Conversely if the $*$-action of G is transitive then $\pi(e_0)$ has strictly positive entries, so by Theorem 1, π is irreducible. ∎

4. The Basic Inequality

In this section we will prove the following result and some related facts.

Theorem 5. *Suppose a commutative hypergroup $\mathcal{K}$ acts irreducibly on a finite set X. Then*

$$|X| \leq w(\mathcal{K}).$$

We will use some basic facts about harmonic analysis on a commutative hypergroup $\mathcal{K} = \{c_0, c_1, \ldots, c_n\} \subset \mathbb{A}$ from [13] which we now review. There exists a basis $\{e_0, e_1, \ldots, e_n\}$ of $\mathcal{A}$ consisting of mutually orthogonal idempotents (this means $e_i e_j = \delta_{i,j} e_i$) such that

$$c_i e_j = \chi_j(c_i) e_j \text{ for all } c_i \in \mathcal{K}$$

for some complex-valued functions χ_j. These functions are exactly the *characters* of $\mathcal{K}$, that is, those functions χ such that

(i) $\chi(c_i)\chi(c_j) = \sum_k n_{ij}^k \chi(c_k)$
(ii) $\chi(c_i^*) = \overline{\chi(c_i)}$.

The set $\{\chi_0, \chi_1, \ldots, \chi_n\}$ of characters of $\mathcal{K}$ is denoted by $\mathcal{K}^\wedge$. Characters are orthogonal with respect to the inner product

$$\langle f, g \rangle = \frac{1}{w(\mathcal{K})} \sum_{i=0}^{n} w(c_i) f(c_i) \overline{g(c_i)}$$

and form a signed hypergroup under pointwise multiplication and complex conjugation.

The weights $w(\chi_j)$ are still well-defined and positive, and $w(\mathcal{K}^\wedge) = \sum_j w(\chi_j) = w(\mathcal{K})$. Furthermore we have explicit formulae for the relationships between the bases $\{c_0, \ldots, c_n\}$ and $\{e_0, \ldots, e_n\}$ of $\mathcal{A}$:

$$\begin{aligned} e_j &= \frac{w(\chi_j)}{w(\mathcal{K})} \sum_{i=0}^{n} w(c_i) \chi_j(c_i) c_i \\ c_i &= \sum_j \chi_j(c_i) e_j. \end{aligned}$$

In particular $\chi_0 \equiv 1$ and e_0 is as given previously.

Assume we now have a $*$-action of the commutative hypergroup $\mathcal{K} \subset \mathcal{A}$ on a set $X = \{x_1, \ldots, x_k\}$. Let $\mathcal{B}$ denote the complex vector space with basis X; that is $\mathcal{B} = \{\sum_{i=1}^{k} z_i x_i \mid z_i \in \mathbb{C}\}$. The $*$-action π of $\mathcal{K}$ on X extends linearly to a $*$-action of $\mathcal{A}$ on the vector space $\mathcal{B}$. Define $\mathcal{B}_j = \pi(e_j)\mathcal{B}$ and let dim $\mathcal{B}_j = b_j$.

Definition 14. The vector $\mathbf{b} = [b_0, \ldots, b_n]$ will be called the *multiplicity vector* of the action.

Since the e_j are orthogonal idempotents and $c_0 = \sum_j e_j$, we get

$$\mathcal{B} = \oplus_{j=0}^{n} \mathcal{B}_j$$

where $\pi(e_j)$ is the projection onto $\mathcal{B}_j$. Then

$$b_j = \text{tr } \pi(e_j) = \frac{w(\chi_j)}{w(\mathcal{K})} \sum_{i=0}^{n} w(c_i)\chi_j(c_i) \text{ tr } \pi(c_i).$$

All the quantities involved in the right hand side of this expression are positive, except for $\chi_j(c_i)$, which is a complex number of modulus at most one.

Thus we have

$$\begin{aligned} b_j &\leq \frac{w(\chi_j)}{w(\mathcal{K})} \sum_{i=0}^{n} w(c_i) \text{ tr } \pi(c_i) \\ &= w(\chi_j) \text{ tr } \pi(e_0) \\ &= w(\chi_j) \end{aligned}$$

since by assumption π is irreducible and so $\pi(e_0)$ is a rank one projection and has trace 1. Then

$$|X| = \sum_{j=0}^{n} b_j \leq \sum_{j=0}^{n} w(\chi_j) = w(\mathcal{K}^{\wedge}) = w(\mathcal{K}).$$

This proves our theorem and in addition establishes the following.

Proposition 6. *(a) If* $b_j =$ dim $\mathcal{B}_j =$ dim $\pi(e_j)\mathcal{B}$ *then*

$$b_j \leq w(\chi_j).$$

(b) The character of the action is given by

$$\chi(c_i) = \sum_{j=0}^{n} b_j \chi_j(c_i).$$

Definition 16. An action π of a hypergroup $\mathcal{K}$ on a set X is *maximal* if $|X| = w(\mathcal{K})$.

For a maximal action to exist, $w(\mathcal{K})$ must be an integer. Maximal $*$-actions, if they exist, are of particular interest due to a connection with association schemes. We define these combinatorial objects here; for more information see [1], [3], [5].

Definition 17. An association scheme on a finite set of k elements is a set $\{A_0, A_1, \ldots, A_n\}$ of $k \times k$ matrices with the following properties:

(i) Each entry of any A_i is either 0 or 1.

(ii) A_0 is the identity matrix.

(iii) There exist non-negative integers $p_{i,j}^l$ such that $A_i A_j = \sum_{l=0}^n p_{i,j}^l A_l$.

(iv) The set $\{A_0, A_1, \ldots, A_n\}$ is closed under formation of adjoints

(v) $\sum_{l=0}^n A_l = J_k$.

If A_{i^*} is the adjoint of A_i, then the number $w_i = p_{i*,i}^0$ is called the valency of the 'i-th' class of the association scheme. It is not hard to show that if we define $c_i = w_i^{-1} A_i$, then $\{c_0, c_1, \ldots, c_n\}$ forms a hypergroup of matrices.

Conversely, given a hypergroup $\mathcal{K}$, when does it arise in this fashion from some association scheme? The answer is given, at least partially, by the following (for a proof see [12]).

Theorem 7. *(a) Suppose that a hypergroup $\mathcal{K}$ admits a maximal $*$-action $\pi : \mathcal{K} \to Aff(X)$. Then the matrices $\{A_i = w(c_i) T_{\pi(c_i)}, 0 \leq i \leq n\}$ define an association scheme, and $\mathcal{K}$ arises from that association scheme in the manner described above.*

(b) Conversely if a hypergroup $\mathcal{K}$ arises from an association scheme in the manner described above, then $\mathcal{K}$ admits a maximal $*$-action.

5. Some specific examples

In the examples that follow, we will not distinguish between an affine map $\pi(c_i)$ and the corresponding matrix $T_{\pi(c_i)}$.

Example 1. Hypergroups of order two

A hypergroup $\mathcal{K} = \{c_0, c_1\}$ with two elements is determined by the single equation

$$c_1^2 = \alpha c_0 + (1-\alpha)c_1$$

where $0 < \alpha \leq 1$ is arbitrary. Then $\omega(\mathcal{K}) = (\alpha+1)/\alpha$ and the Haar measure is

$$e_0 = \frac{\alpha}{\alpha+1}c_0 + \frac{1}{\alpha+1}c_1.$$

Suppose that $\mathcal{K}$ $*$-acts irreducibly on a set X with $|X| = k$. By Corollary 9,

$$\pi(e_0) = \frac{1}{k}J_k$$

and so

$$\pi(c_1) = (\alpha+1)\pi(e_0) - \alpha\pi(c_0) = \frac{\alpha+1}{k}J_k - \alpha I_k.$$

Provided that $\frac{\alpha+1}{k} \geq \alpha$ (the inequality of Theorem 13 in this special case), this is a doubly stochastic matrix.

Note that

$$\begin{aligned}
\pi(c_1)^2 &= \frac{(\alpha+1)^2}{k^2}J_k^2 - 2\frac{\alpha(\alpha+1)}{k}J_k + \alpha^2 I_k \\
&= \frac{(\alpha+1)(1-\alpha)}{k}J_k + \alpha^2 I_k \\
&= \alpha I_k + (1-\alpha)\left(\frac{(\alpha+1)}{k}J_k - \alpha I_k\right) \\
&= \alpha\pi(c_0) + (1-\alpha)\pi(c_1)
\end{aligned}$$

so this is indeed a $*$-action. Summarizing, we have

Proposition 8. *The hypergroup $\mathcal{K} = \{c_0, c_1\}$ with structure equation $c_1^2 = \alpha c_0 + (1-\alpha)c_1$ for $\alpha \in (0,1]$ has exactly one irreducible $*$-action on any set X of size $k \leq \frac{\alpha+1}{\alpha}$. This $*$-action is given by*

$$\pi(c_1) = \frac{\alpha+1}{k}J_k - \alpha I_k.$$

Example 2. $\mathcal{K}(S_3)$

The class hypergroup of the symmetric group S_3 is $\mathcal{K}(S_3) = \{c_0, c_1, c_2\}$ with structure equations

$$c_1^2 = \frac{1}{3}c_0 + \frac{2}{3}c_2$$

From Theorem 13 we know that $\mathcal{K}(S_3)$ can $*$-act irreducibly only a set X with $|X| \leq 6$. From Theorem 12 we also know that there are irreducible $*$-actions of dimensions $k = 1, 2, 3$ and 6. Are there any others?

First we note that $\mathcal{L} = \{c_0, c_2\}$ is a subhypergroup of $\mathcal{K}$ of total weight 3, so that any irreducible $*$-action of $\mathcal{K}$ decomposes upon restriction to $\mathcal{L}$ into irreducibles of size 1, 2 or 3.

Using Proposition 18, we find the irreducible 2 and 3 dimensional $*$-actions of $\mathcal{L}$ are given by one of the following

$$\pi(c_2) = \frac{1}{4}\begin{bmatrix} 1 & 3 \\ 3 & 1 \end{bmatrix} \quad \text{or} \quad \pi(c_2) = \frac{1}{2}\begin{bmatrix} 0 & 1 & 1 \\ 1 & 0 & 1 \\ 1 & 1 & 0 \end{bmatrix}.$$

Note that if $|X| = k$ then

$$\begin{aligned} \pi(c_1) &\doteq 2\pi(e_0) - \frac{1}{3}\pi(c_0) - \frac{2}{3}\pi(c_2) \\ &= \frac{2}{k}J_k - \frac{1}{3}I_k - \frac{2}{3}\pi(c_2) \end{aligned}$$

and is determined by $\pi(c_2)$. This equation also shows that $\pi\mid_{\mathcal{L}}$ cannot contain any 1 dimensional $*$-actions unless $k = 2$, as otherwise a negative term would appear somewhere on the diagonal of $\pi(c_1)$.

If $k = 2$, there are 2 possibilities according as $\pi\mid_{\mathcal{L}}$ is reducible or irreducible. These are:

$$\pi(c_1) = \begin{bmatrix} 0 & 1 \\ 1 & 0 \end{bmatrix} \quad \text{and} \quad \pi(c_2) = \begin{bmatrix} 1 & 0 \\ 0 & 1 \end{bmatrix}$$

or

$$\pi(c_1) = \frac{1}{2}\begin{bmatrix} 1 & 1 \\ 1 & 1 \end{bmatrix} \quad \text{and} \quad \pi(c_2) = \frac{1}{4}\begin{bmatrix} 1 & 3 \\ 3 & 1 \end{bmatrix}.$$

The first of these corresponds to the $*$-action on G/H, where $|H| = 3$.

If $k = 3$, $\pi\mid_{\mathcal{L}}$ must be irreducible so we get

$$\pi(c_1) = \frac{1}{3}\begin{bmatrix} 1 & 1 & 1 \\ 1 & 1 & 1 \\ 1 & 1 & 1 \end{bmatrix} \quad \text{and} \quad \pi(c_2) = \frac{1}{2}\begin{bmatrix} 0 & 1 & 1 \\ 1 & 0 & 1 \\ 1 & 1 & 0 \end{bmatrix}$$

which corresponds to the $*$-action on G/H, where $|H| = 2$.

If $k = 4$, $\pi\,|_{\mathcal{L}}$ must split into 2 irreducibles of dimension 2, so again $\pi(c_2)$ is determined and we get

$$\pi(c_1) = \frac{1}{2}\begin{bmatrix} 0 & 0 & 1 & 1 \\ 0 & 0 & 1 & 1 \\ 1 & 1 & 0 & 0 \\ 1 & 1 & 0 & 0 \end{bmatrix} \quad \text{and} \quad \pi(c_2) = \frac{1}{4}\begin{bmatrix} 1 & 3 & 0 & 0 \\ 3 & 1 & 0 & 0 \\ 0 & 0 & 1 & 3 \\ 0 & 0 & 3 & 1 \end{bmatrix}.$$

If $k = 5$, $\pi\,|_{\mathcal{L}}$ must split into irreducibles of dimension 2 and 3. But then $\pi(c_2)$ has two entries of $\frac{1}{4}$ on the diagonal (thanks to the dimension 2 component) and the corresponding entries of $\pi(c_1)$ will be $\frac{2}{5}-\frac{1}{3}-\frac{2}{3}\times\frac{1}{4} < 0$ which is impossible. Therefore no dimension 5 $*$-action exists.

If $k = 6$, $\pi\,|_{\mathcal{L}}$ must split into 2 pieces of dimension 3 by the above argument for $k = 5$. We obtain

$$\pi(c_1) = \frac{1}{3}\begin{bmatrix} 0 & 0 & 0 & 1 & 1 & 1 \\ 0 & 0 & 0 & 1 & 1 & 1 \\ 0 & 0 & 0 & 1 & 1 & 1 \\ 1 & 1 & 1 & 0 & 0 & 0 \\ 1 & 1 & 1 & 0 & 0 & 0 \\ 1 & 1 & 1 & 0 & 0 & 0 \end{bmatrix} \quad \text{and} \quad \pi(c_2) = \frac{1}{2}\begin{bmatrix} 0 & 1 & 1 & 0 & 0 & 0 \\ 1 & 0 & 0 & 0 & 0 & 0 \\ 1 & 1 & 0 & 0 & 0 & 0 \\ 0 & 0 & 1 & 0 & 1 & 1 \\ 0 & 0 & 0 & 1 & 0 & 1 \\ 0 & 0 & 0 & 1 & 1 & 0 \end{bmatrix}.$$

This corresponds to the $*$-action of $\mathcal{K}$ on G itself– a maximal $*$-action which corresponds to the association scheme given by the conjugacy classes of G. Summarizing, we have the following:

Proposition 20. *Let* $\mathcal{K} = \{c_0, c_1, c_2\} = \mathcal{K}(S_3)$ *with structure equations* $c_1^2 = \frac{1}{3}c_0 + \frac{2}{3}c_2$, $c_1c_2 = c_1$ *and* $c_2^2 = \frac{1}{2}c_0 + \frac{1}{2}c_2$. *Then* $\mathcal{K}$ *has exactly the following irreducible* $*$*-actions:*

(i) $\pi(c_1) = \pi(c_2) = [1]$

(ii) $\pi(c_1) = \begin{bmatrix} 0 & 1 \\ 1 & 0 \end{bmatrix}$ *and* $\pi(c_2) = \begin{bmatrix} 1 & 0 \\ 0 & 1 \end{bmatrix}$

(iii) $\pi(c_1) = \frac{1}{2}\begin{bmatrix} 1 & 1 \\ 1 & 1 \end{bmatrix}$ *and* $\pi(c_2) = \frac{1}{4}\begin{bmatrix} 1 & 3 \\ 3 & 1 \end{bmatrix}$

(iv) $\pi(c_1) = \frac{1}{3}\begin{bmatrix} 1 & 1 & 1 \\ 1 & 1 & 1 \\ 1 & 1 & 1 \end{bmatrix}$ *and* $\pi(c_2) = \frac{1}{2}\begin{bmatrix} 0 & 1 & 1 \\ 1 & 0 & 1 \\ 1 & 1 & 0 \end{bmatrix}$

(v) $\pi(c_1) = \frac{1}{2}\begin{bmatrix} 0 & 0 & 1 & 1 \\ 0 & 0 & 1 & 1 \\ 1 & 1 & 0 & 0 \\ 1 & 1 & 0 & 0 \end{bmatrix}$ *and* $\pi(c_2) = \frac{1}{4}\begin{bmatrix} 1 & 3 & 0 & 0 \\ 3 & 1 & 0 & 0 \\ 0 & 0 & 1 & 3 \\ 0 & 0 & 3 & 1 \end{bmatrix}$

$$(vi)\ \pi(c_1) = \frac{1}{3}\begin{bmatrix} 0 & 0 & 0 & 1 & 1 & 1 \\ 0 & 0 & 0 & 1 & 1 & 1 \\ 0 & 0 & 0 & 1 & 1 & 1 \\ 1 & 1 & 1 & 0 & 0 & 0 \\ 1 & 1 & 1 & 0 & 0 & 0 \\ 1 & 1 & 1 & 0 & 0 & 0 \end{bmatrix} \ and\ \pi(c_2) = \frac{1}{2}\begin{bmatrix} 0 & 1 & 1 & 0 & 0 & 0 \\ 1 & 0 & 1 & 0 & 0 & 0 \\ 1 & 1 & 0 & 0 & 0 & 0 \\ 0 & 0 & 0 & 0 & 1 & 1 \\ 0 & 0 & 0 & 1 & 0 & 1 \\ 0 & 0 & 0 & 1 & 1 & 0 \end{bmatrix}$$

We find that besides the 'expected' $*$-actions on the homogeneous spaces of G, we also have two others of dimensions 2 and 4.

Example 3. $\mathcal{K}(S_3^\wedge)$

For our next example, we consider the character hypergroup of S_3 given by $\mathcal{K}(S_3^\wedge) = \{\chi_0, \chi_1, \chi_2\}$ with equations

$$\begin{aligned} \chi_1^2 &= \chi_0 \\ \chi_1\chi_2 &= \chi_2 \\ \chi_2^2 &= \frac{1}{4}\chi_0 + \frac{1}{4}\chi_1 + \frac{1}{2}\chi_2 \end{aligned}$$

Then $w(\chi_1) = 1$ and $w(\chi_2) = 4$ so that

$$e_0 = \frac{1}{6}\chi_0 + \frac{1}{6}\chi_1 + \frac{2}{3}\chi_2.$$

$\mathcal{L} = \{\chi_0, \chi_1\}$ is a subhypergroup of total weight 2 whose only nontrivial irreducible $*$-action is given by

$$\pi(\chi_1) = \begin{bmatrix} 0 & 1 \\ 1 & 0 \end{bmatrix}.$$

The restriction of any $*$-action to $\mathcal{L}$ is a sum of trivial and 2-dimensional irreducibles. Since from Corollary 9 we have

$$\pi(\chi_2) = \frac{3}{2}\frac{J_k}{k} - \frac{1}{4}I_k - \frac{1}{4}\pi(\chi_1)$$

we may check the possibilities for $\pi(\chi_1)$ which are consistent with the condition $\pi(\chi_2) \in \Omega_k$.

For $k = 2$, the possibilities are

$$\pi(\chi_1) = \begin{bmatrix} 1 & 0 \\ 0 & 1 \end{bmatrix} \text{ and } \pi(\chi_2) = \frac{1}{4}\begin{bmatrix} 1 & 3 \\ 3 & 1 \end{bmatrix}$$

or

$$\pi(\chi_1) = \begin{bmatrix} 0 & 1 \\ 1 & 0 \end{bmatrix} \text{ and } \pi(\chi_2) = \frac{1}{2}\begin{bmatrix} 1 & 1 \\ 1 & 1 \end{bmatrix}.$$

For $k = 3$ the possibilities are

$$\pi(\chi_1) = \begin{bmatrix} 1 & 0 & 0 \\ 0 & 1 & 0 \\ 0 & 0 & 1 \end{bmatrix} \text{ and } \pi(\chi_2) = \frac{1}{2}\begin{bmatrix} 0 & 1 & 1 \\ 1 & 0 & 1 \\ 1 & 1 & 0 \end{bmatrix}$$

or

$$\pi(\chi_1) = \begin{bmatrix} 1 & 0 & 0 \\ 0 & 0 & 1 \\ 0 & 1 & 0 \end{bmatrix} \text{ and } \quad \pi(\chi_2) = \frac{1}{4}\begin{bmatrix} 0 & 2 & 2 \\ 2 & 1 & 1 \\ 2 & 1 & 1 \end{bmatrix}.$$

For $k \geq 4$, $\pi \mid \mathcal{L}$ cannot contain any 1 dimensional $*$-actions since then by $(*)$, a diagonal term of $\pi(\chi_1)$ would be

$$\frac{3}{2k} - \frac{1}{4} - \frac{1}{4} < 0.$$

Thus for $k = 4$ we get

$$\pi(\chi_1) = \begin{bmatrix} 0 & 1 & 0 & 0 \\ 1 & 0 & 0 & 0 \\ 0 & 0 & 0 & 1 \\ 0 & 0 & 1 & 0 \end{bmatrix} \text{ and } \pi(\chi_2) = \frac{1}{8}\begin{bmatrix} 1 & 1 & 3 & 3 \\ 1 & 1 & 3 & 3 \\ 3 & 3 & 1 & 1 \\ 3 & 3 & 1 & 1 \end{bmatrix}$$

and for $k = 5$ there are no possibilities.

For $k = 6$ we get

$$\pi(\chi_1) = \begin{bmatrix} 0 & 1 & 0 & 0 & 0 & 0 \\ 1 & 0 & 0 & 0 & 0 & 0 \\ 0 & 0 & 0 & 1 & 0 & 0 \\ 0 & 0 & 1 & 0 & 0 & 0 \\ 0 & 0 & 0 & 0 & 0 & 1 \\ 0 & 0 & 0 & 0 & 1 & 0 \end{bmatrix} \text{ and } \pi(\chi_2) = \frac{1}{4}\begin{bmatrix} 0 & 0 & 1 & 1 & 1 & 1 \\ 0 & 0 & 1 & 1 & 1 & 1 \\ 1 & 1 & 0 & 0 & 1 & 1 \\ 1 & 1 & 0 & 0 & 1 & 1 \\ 1 & 1 & 1 & 1 & 0 & 0 \\ 1 & 1 & 1 & 1 & 0 & 0 \end{bmatrix}$$

Summarizing, we have the following

Proposition 21. *Let $\mathcal{K} = \{\chi_0, \chi_1, \chi_2\} = \mathcal{K}(S_3^\wedge)$ with structure equations $\chi_2^1 = \chi_0$, $\chi_1\chi_2 = \chi_2$ and $\chi_2^2 = \frac{1}{4}\chi_0 + \frac{1}{4}\chi_1 + \frac{1}{2}\chi_2$. Then $\mathcal{K}$ has exactly the following irreducible $*$-actions:*

(i) $\pi(\chi_1) = \pi(\chi_2) = [1]$

(ii) $\pi(\chi_1) = \begin{bmatrix} 1 & 0 \\ 0 & 1 \end{bmatrix}$ *and* $\pi(\chi_2) = \frac{1}{4}\begin{bmatrix} 1 & 3 \\ 3 & 1 \end{bmatrix}$

(iii) $\pi(\chi_1) = \begin{bmatrix} 0 & 1 \\ 1 & 0 \end{bmatrix}$ *and* $\pi(\chi_2) = \frac{1}{2}\begin{bmatrix} 1 & 1 \\ 1 & 1 \end{bmatrix}$

$$(iv)\ \pi(\chi_1) = \begin{bmatrix} 1 & 0 & 0 \\ 0 & 1 & 0 \\ 0 & 0 & 1 \end{bmatrix} \text{ and } \pi(\chi_2) = \tfrac{1}{2}\begin{bmatrix} 0 & 1 & 1 \\ 1 & 0 & 1 \\ 1 & 1 & 0 \end{bmatrix}$$

$$(v)\ \pi(\chi_1) = \begin{bmatrix} 1 & 0 & 0 \\ 0 & 0 & 1 \\ 0 & 1 & 0 \end{bmatrix} \text{ and } \pi(\chi_2) = \tfrac{1}{4}\begin{bmatrix} 0 & 2 & 2 \\ 2 & 1 & 1 \\ 2 & 1 & 1 \end{bmatrix}$$

$$(vi)\ \pi(\chi_1) = \begin{bmatrix} 0 & 1 & 0 & 0 \\ 1 & 0 & 0 & 0 \\ 0 & 0 & 0 & 1 \\ 0 & 0 & 1 & 0 \end{bmatrix} \text{ and } \pi(\chi_2) = \tfrac{1}{8}\begin{bmatrix} 1 & 1 & 3 & 3 \\ 1 & 1 & 3 & 3 \\ 3 & 3 & 1 & 1 \\ 3 & 3 & 1 & 1 \end{bmatrix}$$

$$(vii)\ \pi(\chi_1) = \begin{bmatrix} 0 & 1 & 0 & 0 & 0 & 0 \\ 1 & 0 & 0 & 0 & 0 & 0 \\ 0 & 0 & 0 & 1 & 0 & 0 \\ 0 & 0 & 1 & 0 & 0 & 0 \\ 0 & 0 & 0 & 0 & 0 & 1 \\ 0 & 0 & 0 & 0 & 1 & 0 \end{bmatrix} \text{ and } \pi(\chi_2) = \tfrac{1}{4}\begin{bmatrix} 0 & 0 & 1 & 1 & 1 & 1 \\ 0 & 0 & 1 & 1 & 1 & 1 \\ 1 & 1 & 0 & 0 & 1 & 1 \\ 1 & 1 & 0 & 0 & 1 & 1 \\ 1 & 1 & 1 & 1 & 0 & 0 \\ 1 & 1 & 1 & 1 & 0 & 0 \end{bmatrix}.$$

Example 4. The Golden hypergroup

The Golden hypergroup arises from the association scheme of a pentagon viewed as a graph; by this we mean that the two non trivial classes are given by the relations of being neighbours or not being neighbours respectively. $K = \{c_0, c_1, c_2\}$ has structure equations

$$\begin{aligned} c_1^2 &= \frac{1}{2}c_0 + \frac{1}{2}c_2 \\ c_1c_2 &= \frac{1}{2}c_1 + \frac{1}{2}c_2 \\ c_2^2 &= \frac{1}{2}c_0 + \frac{1}{2}c_1 \end{aligned}$$

and character table

	c_0	c_1	c_2
χ_0	1	1	1
χ_1	1	$\frac{-1+\sqrt{5}}{4}$	$\frac{-1-\sqrt{5}}{4}$
χ_2	1	$\frac{-1-\sqrt{5}}{4}$	$\frac{-1+\sqrt{5}}{4}$

Note that from the table we see that $\mathcal{K}^\wedge \simeq \mathcal{K}$ and so that $w(\chi_1 = w(\chi_2) = 2$. We know from Theorem 13 that any irreducible *-action must be of dimension $k \leq w(K) = 5$.

If $k = 2$ there are two possibilities for the dimension vector; $\mathbf{b} = [1,1,0]$ or $\mathbf{b} = [1,0,1]$ which from the character table results in the following two possibilities for the character; $\chi = (2, \frac{3+\sqrt{5}}{4}, \frac{3-\sqrt{5}}{4})$ or $\chi = (2, \frac{3-\sqrt{5}}{4}, \frac{3+\sqrt{5}}{4})$. Since a two dimensional doubly stochastic matrix is determined by its trace, we get the following possibilities:

$$\pi(c_1) = 1/8 \begin{bmatrix} 3+\sqrt{5} & 5-\sqrt{5} \\ 5-\sqrt{5} & 3+\sqrt{5} \end{bmatrix}, \; \pi(c_2) = 1/8 \begin{bmatrix} 3-\sqrt{5} & 5+\sqrt{5} \\ 5+\sqrt{5} & 3-\sqrt{5} \end{bmatrix},$$

$$\pi(c_1) = 1/8 \begin{bmatrix} 3-\sqrt{5} & 5+\sqrt{5} \\ 5+\sqrt{5} & 3-\sqrt{5} \end{bmatrix}, \pi(c_2) = 1/8 \begin{bmatrix} 3+\sqrt{5} & 5-\sqrt{5} \\ 5-\sqrt{5} & 3+\sqrt{5} \end{bmatrix}$$

Both of these can be easily checked to be $*$-actions.

If $k = 3$ the situation is more subtle. We will show that no $*$-action π exists by assuming the contrary. From the character table the only possibility for the multiplicity vector is $\mathbf{b} = [1,1,1]$. Suppose that we place the simplex X on which $\mathcal{K}$ acts in a Euclidean plane as an equilateral triangle whose vertices are on the circle of radius one about the origin O. Each of $T = \pi(c_1)$, $U = \pi(c_2)$ are then real symmetric linear transformations of this plane onto itself which commute and whose eigenvalues are known from the character table and the multiplicity vector. This means we may choose orthogonal coordinates (x, y) such that if r and s denote the quantities $\frac{-1+\sqrt{5}}{4}$ and $\frac{-1-\sqrt{5}}{4}$ respectively then T and U are the linear transformations $T(x,y) = (rx, sy)$ and $U(x,y) = (sx, ry)$.

By assumption both T and U map the equilateral triangle into itself. In particular since the origin divides any median of this triangle in the ratio $2:1$, for any vertex $v = (\cos\theta, \sin\theta)$ we must have $\langle Tv, v\rangle \geq -\frac{1}{2}$ and $\langle Uv, v\rangle \geq -\frac{1}{2}$. Thus

$$\begin{aligned} r\cos^2\theta + s\sin^2\theta &\geq -\frac{1}{2} \\ s\cos^2\theta + r\sin^2\theta &\geq -\frac{1}{2}. \end{aligned}$$

The second of these equations may be manipulated easily to get

$$\cos^2\theta \leq \frac{1+\sqrt{5}}{2\sqrt{5}}$$

Now $\frac{1+\sqrt{5}}{2\sqrt{5}} \leq \cos^2\pi/6$ so that $|\cos\theta| \leq \cos\pi/6 = \sqrt{3}/2$. Similiarly $|\sin\theta| \leq \sqrt{3}/2$. This means that θ can only lie in the range $\pi/6 \leq \theta \leq \pi/3 \pmod{\pi/2}$. But then one of $\theta + 2\pi/3$, $\theta - 2\pi/3$ lies outside this range mod $\pi/2$. But these are the other vertices of the equilateral triangle we are considering so the same argument must apply to them.

This contradiction shows that no irreducible $*$-action of dimension $k = 3$ exists.

If $k = 4$ the possibilities for the multiplicity vector $\mathbf{b} = [1,3,0]$ or $\mathbf{b} = [1,2,1]$ are easily dismissed; the first by Proposition 15 and the second by the fact that the corresponding character would be $\chi = (4, \frac{1+\sqrt{5}}{4}, \frac{1-\sqrt{5}}{4})$. The same is true for the alternatives $\mathbf{b} = [1,0,3]$ and $\mathbf{b} = [1,1,2]$.

If $k = 5$ then the multiplicity vector is necessarily $\mathbf{b} = [1,2,2]$ and the associated character is $\chi = [5,0,0]$. By Theorem 7 we know that $2\pi(c_1)$ and $2\pi(c_2)$ are $0,1$ symmetric matrices which sum to $J_5 - I_5$. This means that each must have exactly two 1's in each row (and column). A bit of thought shows that up to isomorphism there is only one such arrangement and that it does indeed define a $*$-action.

$$\pi(c_1) = \frac{1}{2}\begin{bmatrix} 0 & 1 & 0 & 0 & 1 \\ 1 & 0 & 1 & 0 & 0 \\ 0 & 1 & 0 & 1 & 0 \\ 0 & 0 & 1 & 0 & 1 \\ 1 & 0 & 0 & 1 & 0 \end{bmatrix} \text{ and } \pi(c_2) = \frac{1}{2}\begin{bmatrix} 1 & 0 & 1 & 1 & 0 \\ 0 & 1 & 0 & 1 & 1 \\ 1 & 0 & 1 & 0 & 1 \\ 1 & 1 & 0 & 1 & 0 \\ 0 & 1 & 1 & 0 & 1 \end{bmatrix}$$

This is the association scheme associated to the graph of a pentagon.

References

[1] E. Bannai and T. Ito, *Algebraic Combinatorics I-Association schemes*, Benjamin and Cummings, Menlo Park, 1984.

[2] W. R. Bloom and H. Heyer, *Harmonic Analysis of Probability Measures on Hypergroups*, De Gruyter, Berlin, (1995).

[3] A. E. Brouwer, A. M. Cohen and A. Neumaier, *Distance- Regular Graphs*, Ergeb. der Math., no 3. Band 18, Springer- Verlag, Berlin, 1989.

[4] C. F. Dunkl, *The measure algebra of a locally compact hypergroup*, Trans. Amer. Math. Soc., **179** (1973), 331–348.

[5] C. D. Godsil, *Algebraic Combinatorics*, Chapman and Hall, London, 1993.

[6] R. I. Jewett, *Spaces with an abstract convolution of measures* Adv. Math., **18** (1975), 1–101.

[7] G. L. Litvinov, *Hypergroups and hypergroup algebras*,J. Soviet. Math. **38** (1987), 1734–1761.

[8] K. A. Ross, *Hypergroups and centers of measure algebras*, Symposia Math. **22** (1977), 189–203.

[9] R. Spector, *Mesures invariantes sur les hypergroupes*, Trans. Amer. Math. Soc., **239** (1978), 147–165.

[10] V. S. Sunder, *On the relation between subfactors and hypergroups*, Applications of hypergroups and related measure algebras, Contemp. Math., **183** (1995), 331–340.

[11] V. S. Sunder and A. K. Vijayarajan, *On the non-occurrence of the Coxeter graphs* β_{2n+1}, D_{2n+1} *and* E_7 *as the principal graph of an inclusion of* II_1 *factors*, preprint.

[12] V. S. Sunder and N. J. Wildberger, *On discrete hypergroups and their actions on sets*, preprint (1996).

[13] N. J. Wildberger, *Duality and Entropy for finite commutative hypergroups and fusion rule algebras*, to appear.

[14] N. J. Wildberger, *Lagrange's theorem and integrality for finite commutative hypergroups with applications to strongly-regular graphs*, to appear, J. of Alg.

Institute of Mathematical Sciences, Madras 600113, INDIA

School of Mathematics, University of New South Wales, Sydney, 2052, AUSTRALIA

Positivity of Turán Determinants for Orthogonal Polynomials

*Ryszard Szwarc**

Abstract

The orthogonal polynomials p_n satisfy Turán's inequality if $p_n^2(x) - p_{n-1}(x)p_{n+1}(x) \geq 0$ for $n \geq 1$ and for all x in the interval of orthogonality. We give general criteria for orthogonal polynomials to satisfy Turán's inequality. This yields the known results for classical orthogonal polynomials as well as new results, for example, for the q–ultraspherical polynomials.

1. Introduction

In the 1940s, while studying the zeros of Legendre polynomials $P_n(x)$, Turán [T] discovered that

$$P_n^2(x) - P_{n-1}(x)P_{n+1}(x) \geq 0, \qquad -1 \leq x \leq 1 \tag{1}$$

with equality only for $x = \pm 1$. Szegö [Sz1] gave four different proofs of (1). Shortly after that, analogous results were obtained for other classical orthogonal polynomials such as ultraspherical polynomials [Sk, S], Laguerre and Hermite polynomials [MN], and Bessel functions [Sk, S].

In [KS] Karlin and Szegö raised the question of determining the range of parameters (α, β) for which (1) holds for Jacobi polynomials of order (α, β); i.e. denoting $R_n^{(\alpha,\beta)}(x) = P_n^{(\alpha,\beta)}(x)/P_n^{(\alpha,\beta)}(1)$,

$$[R_n^{(\alpha,\beta)}(x)]^2 - R_{n-1}^{(\alpha,\beta)}(x)R_{n+1}^{(\alpha,\beta)}(x) \geq 0, \qquad -1 \leq x \leq 1. \tag{2}$$

In 1962 Szegö [Sz2] proved (2) for $\beta \geq |\alpha|$, $\alpha > -1$. In a series of two papers [G1, G2] Gasper extended Szegö's result by showing that (2) holds if and only if $\beta \geq \alpha > -1$.

*This work has been partially supported by KBN (Poland) under grant 2 P03A 030 09.

1991 *Mathematics Subject Classification.* Primary 42C05, 47B39

Key words and phrases: orthogonal polynomials, Turán's inequality, recurrence formula.

More recently, attention has also turned to the q-analogues of the classical polynomials [BI1].

All the results mentioned above were proved using differential equations, that the classical orthogonal polynomials satisfy. Therefore the methods cannot be used to extend (1) to more general orthogonal polynomials. In 1970 Askey [A, Thm. 3] gave a general criterion for monic symmetric orthogonal polynomials to satisfy the Turán type inequality on the entire real line. His result, however, does not imply (1) for the Legendre polynomials because the latter are not monic in the standard normalization, and they do not satisfy Askey's assumptions in the monic normalization. In this paper we give general criteria for orthogonal polynomials implying (1) holds for x in the support of corresponding orthogonality measure. The assumptions are stated in terms of the coefficients of the recurrence relation that the orthogonal polynomials satisfy. They admit a very simple form in the case of symmetric orthogonal polynomials; i.e. the case $p_n(-x) = (-1)^n p_n(x)$. In particular, the results apply to all the ultraspherical polynomials, giving yet another proof of Turán's inequality for the Legendre polynomials.

It turns out that the way we normalize the polynomials is essential for the Turán inequality to hold. The results concerning the classical orthogonal polynomials used the normalization at one endpoint of the interval of orthogonality, e.g. at $x = 1$ for the Jacobi polynomials and at $x = 0$ for the Laguerre polynomials. We will also use this normalization and will show that this choice is optimal (Proposition 1). However, the recurrence relation for the polynomials normalized in this way may not be available explicitly. This is the case of the q–ultraspherical polynomials. We give a way of overcoming this obstacle (Corollary 1). In particular, we prove the Turán inequality for all q-ultraspherical polynomials with $q > 0$. These polynomials have been studied by Bustoz and Ismail [BI1] but with a normalization other than at $x = 1$. The same method is applied to the symmetric Pollaczek polynomials, studied in [BI2], again with different normalization.

In Section 6 we prove results for nonsymmetric orthogonal polynomials (Thm. 4). The assumptions again are given in terms of the coefficients in a three term recurrence relation but they are much more involved.

In Section 7 we state results concerning polynomials orthogonal on the positive half axis. In particular they can be applied to the Laguerre polynomials of any order α.

2. Basic formulas

Let p_n be polynomials orthogonal with respect to a probability measure on $\mathbb{R}$. The expressions

$$\Delta_n(x) = p_n^2(x) - p_{n-1}(x)p_{n+1}(x) \quad n = 0, 1, \dots, \tag{3}$$

are called the Turán determinants. Our goal is to give conditions implying the nonnegativity of $\Delta_n(x)$ for x in the support of the orthogonality measure.

The first problem we encounter is that the orthogonality determines the polynomials p_n up to a nonzero multiple. The sign of $\Delta_n(x)$ may change if we multiply each p_n by different nonzero constants. We will normalize the polynomials p_n to obtain the sharpest results possible. Namely, we will assume that

$$p_n(a) = 1$$

at a point a in the support of the orthogonality measure. In this way the Turán determinant vanishes at $x = a$.

Our main interest is focused on the case when the orthogonality measure is supported in a bounded interval. By an affine change of variables we can assume that this interval is $[-1, 1]$. In that case we set $a = 1$. Since the polynomials p_n do not change sign in the interval $[1, +\infty)$ they have positive leading coefficients.

Assume that the polynomials p_n are orthogonal, with positive leading coefficients and $p_n(1) = 1$. Then they satisfy the three term recurrence relation

$$xp_n(x) = \gamma_n p_{n+1}(x) + \beta_n p_n(x) + \alpha_n p_{n-1}(x) \quad n = 0, 1, \dots, \tag{4}$$

with initial conditions $p_{-1} = 0$, $p_0 = 1$, where α_n, β_n, and γ_n are given sequences of real valued coefficients such that

$$\alpha_0 = 0, \quad \alpha_{n+1} > 0, \quad \gamma_n > 0 \quad \text{for } n = 0, 1, \dots.$$

Plugging $x = 1$ into (4) gives

$$\alpha_n + \beta_n + \gamma_n = 1 \qquad n = 0, 1, \dots. \tag{5}$$

Proposition 1. *Let the polynomials p_n satisfy (4) and (5). Then*

$$\gamma_n\Delta_n = \gamma_n p_n^2 + \alpha_n p_{n-1}^2 - (x - \beta_n)p_{n-1}p_n, \tag{6}$$

$$\gamma_n\Delta_n = (p_{n-1} - p_n)[(\gamma_{n-1} - \gamma_n)p_n + (\alpha_n - \alpha_{n-1})p_{n-1}] + \alpha_{n-1}\Delta_{n-1}, \tag{7}$$

$$\gamma_n\Delta_n = (p_n - p_{n-1})(\gamma_n p_n - \alpha_n p_{n-1}) + (1 - x)p_{n-1}p_n, \tag{8}$$

for $n = 1, 2, \dots$.

Proof. By (4) we get

$$\begin{aligned}\gamma_n\Delta_n &= \gamma_n p_n^2 - \gamma_n p_{n-1}[(x-\beta_n) - \alpha_n p_{n-1}] \\ &= \gamma_n p_n^2 + \alpha_n p_{n-1}^2 - (x-\beta_n)p_{n-1}p_n \\ &= \gamma_n p_n^2 + \alpha_n p_{n-1}^2 - (\beta_{n-1}-\beta_n)p_{n-1}p_n - (x-\beta_{n-1})p_{n-1}p_n.\end{aligned}$$

Now applying (4), with n replaced by $n-1$, to the last term yields

$$\gamma_n\Delta_n = (\gamma_n - \gamma_{n-1})p_n^2 + (\alpha_n - \alpha_{n-1})p_{n-1}^2 - (\beta_{n-1}-\beta_n)p_{n-1}p_n + \alpha_{n-1}\Delta_{n-1}.$$

The use of

$$\beta_{n-1} - \beta_n = (\gamma_n - \gamma_{n-1}) + (\alpha_n - \alpha_{n-1})$$

concludes the proof of (7). In order to get (8) replace β_n with $1-\alpha_n-\gamma_n$ in (6). ∎

3. Symmetric polynomials

We will consider first the symmetric orthogonal polynomials, i.e. the orthogonal polynomials satisfying

$$p_n(-x) = (-1)^n p_n(x). \tag{9}$$

Theorem 1. *Let the polynomials p_n satisfy*

$$xp_n(x) = \gamma_n p_{n+1}(x) + \alpha_n p_{n-1}(x) \qquad n = 0,\,1,\,\ldots. \tag{10}$$

with $p_{-1}=0$, $p_0=1$, where $\alpha_0=0$, $\alpha_{n+1}>0$, $\gamma_n>0$, and

$$\alpha_n + \gamma_n = a \qquad n = 0,\,1,\,\ldots.$$

Assume that either (i) or (ii) is satisfied where

(i) α_n is nondecreasing and $\alpha_n \le \frac{a}{2}$ for $n = 1, 2, \ldots$.

(ii) α_n is nonincreasing and $\alpha_n \ge \frac{a}{2}$ for $n = 1, 2, \ldots$.

Then

$$\Delta_n(x) \ge 0, \qquad \textit{for } -a \le x \le a, \quad n = 0,\,1,\,\ldots,$$

and the equality holds if and only if $n \ge 1$ and $x = \pm a$.

Moreover if (i) is satisfied then

$$\Delta_n(x) < 0, \qquad \textit{for } |x| > a, \quad n = 1,\,2,\,\ldots.$$

Proof. By changing variable $x \to ax$ we can restrict ourselves to the case $a = 1$. We prove part (i) only, because the proof of (ii) can be obtained from that of (i) by obvious modifications.

By assumption we have $p_n(1) = 1$ and $p_n(-1) = (-1)^n$. Hence $\Delta_n(\pm 1) = 0$. Assume now that $|x| < 1$. By (9) it suffices to consider $0 \le x < 1$. The proof will go by induction. We have $\gamma_1\Delta_1(x) = \alpha_1(1-x^2) \ge 0$. Now assume $\Delta_{n-1}(x) > 0$.

In view of $\beta_n = 0$ and $\alpha_n - \alpha_{n-1} = \gamma_{n-1} - \gamma_n$, Proposition 1 implies

$$\gamma_n\Delta_n = \gamma_n p_n^2 + \alpha_n p_{n-1}^2 - x p_{n-1} p_n, \tag{11}$$

$$\gamma_n\Delta_n = (\alpha_n - \alpha_{n-1})(p_{n-1}^2 - p_n^2) + \alpha_{n-1}\Delta_{n-1}. \tag{12}$$

By (11) and the positivity of x we may restrict ourselves to the case $p_{n-1}(x)p_n(x) > 0$. We will assume that $p_{n-1}(x) > 0$ and $p_n(x) > 0$ (the case $p_{n-1}(x) < 0$ and $p_n(x) < 0$ can be dealt with similarly). By (12) and by the induction hypothesis it suffices to consider the case $p_{n-1}(x) < p_n(x)$, since by assumption (i) we have $\alpha_{n-1} \le \alpha_n$. In that case since $\gamma_n = 1 - \alpha_n \ge \frac{1}{2} \ge \alpha_n$ we get

$$\gamma_n p_n(x) - \alpha_n p_{n-1}(x) \ge \alpha_n[p_n(x) - p_{n-1}(x)] \ge 0.$$

Now we apply (8) and obtain

$$\gamma_n\Delta_n \ge (1-x)p_{n-1}(x)p_n(x) > 0.$$

The proof of part (i) is thus complete.

We turn to the last part of the statement. Let (i) be satisfied and $|x| > 1$. By symmetry we can assume $x > 1$. As before we proceed by induction. We have

$$\gamma_1\Delta_1(x) = \alpha_1(1-x^2) < 1.$$

Assume now that $\Delta_m(x) < 0$ for $1 \le m \le n-1$. Since $p_n(1) = 1$ and the leading coefficients of p_n's are positive, the polynomials p_n are positive for $x > 1$. Thus

$$0 > \frac{\Delta_m(x)}{p_{m-1}(x)p_m(x)} = \frac{p_m(x)}{p_{m-1}(x)} - \frac{p_{m+1}(x)}{p_m(x)}.$$

for $1 \le m \le n-1$. Hence

$$\frac{p_n(x)}{p_{n-1}(x)} \ge \ldots \ge \frac{p_1(x)}{p_0(x)} = x > 1.$$

Now by (12) we get

$$\gamma_n \Delta_n \leq \alpha_{n-1} \Delta_{n-1} < 0.$$

■

Remark. The second part of Theorem 1 is not true under assumption (ii). Indeed, by (10), the leading coefficient of the Turán determinant $\gamma_n \Delta_n(x)$ is equal to $\gamma_1^{-2} \ldots \gamma_{n-1}^{-2}(\alpha_{n-1} - \alpha_n)$. Thus $\Delta_n(x)$ is positive at infinity for $n \geq 2$. One might expect that in this case $\Delta_n(x)$ is nonnegative on the whole real axis, but this is not true either. Indeed, it can be computed that

$$\gamma_1^2 \gamma_2 \Delta_2(x) = (x^2 - 1)[(\gamma_2 - \gamma_1)x^2 - \alpha_1^2 \gamma_2].$$

One can verify that under assumption (ii) we have

$$r := \frac{\alpha_1^2 \gamma_2}{\gamma_2 - \gamma_1} > 1.$$

(Actually $r \geq 1$ follows from Theorem 1(ii).) Hence $\Delta_2(x) < 0$ for $1 < x < r$.

Sometimes we have to deal with polynomials which are orthogonal in the interval $[-1, 1]$ and normalized at $x = 1$, but the three term recurrence relation is not available in explicit form. In such cases the following will be useful.

Corollary 1 *Let the polynomials p_n satisfy*

$$x p_n = \gamma_n p_{n+1} + \alpha_n p_{n-1}, \qquad n = 0, 1, \ldots,$$

with $p_{-1} = 0$, $p_0 = 1$ and $\alpha_0 = 0$. Assume that the sequences α_n and $\alpha_n + \gamma_n$ are nondecreasing and

$$\lim_{n\to\infty} \alpha_n = \frac{1}{2} a \qquad \lim_{n\to\infty} \gamma_n = \frac{1}{2} a^{-1},$$

where $0 < a < 1$. Then the orthogonality measure for p_n is supported in the interval $[-1, 1]$.

Assume that in addition at least one of the following holds

(i) γ_n is nondecreasing,

(ii) $\gamma_0 \geq 1$.

Then

$$\Delta_n(x) = \tilde{p}_n^2(x) - \tilde{p}_{n-1}(x)\tilde{p}_{n+1}(x) \geq 0 \iff -1 \leq x \leq 1,$$

where $\tilde{p}_n(x) = p_n(x)/p_n(1)$.

Proof. First we will show that $p_n(1) > 0$. In view of symmetry of the polynomials this will imply that the support of the orthogonality measure is contained in $[-1, 1]$.

We will show by induction that $p_n(1)/p_{n-1}(1) \geq a > 0$. We have

$$\frac{p_0(1)}{p_1(1)} = \gamma_0 \leq \frac{1}{2}(a + a^{-1}) \leq a^{-1}.$$

Assume that $p_n(1)/p_{n-1}(1) \geq a$. Then from the recurrence relation we get

$$\frac{p_{n+1}(1)}{p_n(1)} = \frac{1}{\gamma_n}\left(1 - \alpha_n \frac{p_{n-1}(1)}{p_n(1)}\right) \geq \frac{1}{\gamma_n}\left(1 - a^{-1}\alpha_n\right).$$

On the other hand

$$\begin{aligned}
\gamma_n &= (\alpha_n + \gamma_n) - \alpha_n \leq \frac{1}{2}(a + a^{-1}) - \alpha_n \\
&\leq \frac{1}{2}(a + a^{-1}) - a^{-2}\alpha_n + (a^{-2} - 1)\alpha_n \\
&\leq \frac{1}{2}(a + a^{-1}) - a^{-2}\alpha_n + (a^{-2} - 1)\frac{1}{2}a \\
&= a^{-1}(1 - a^{-1}\alpha_n).
\end{aligned}$$

Therefore

$$\frac{p_{n+1}(1)}{p_n(1)} \geq a.$$

Now we show that $c_n = p_n^2(1) - p_{n-1}(1)p_{n+1}(1) > 0$ by induction. Assume (i). Similarly to the proof of Proposition 1 we obtain

$$\gamma_n c_n = (\gamma_n - \gamma_{n-1})p_n^2(1) + (\alpha_n - \alpha_{n-1})p_{n-1}^2 + \alpha_{n-1}c_{n-1}. \tag{13}$$

This implies $c_n > 0$ for every n.

Assume now that (ii) holds and that $c_m > 0$ for $m \leq n - 1$. Hence the sequence $p_{m+1}(1)/p_m(1)$ is positive and nonincreasing for $m \leq n-1$. In particular $p_n(1)/p_{n-1}(1) \leq p_1(1)/p_0(1) = \gamma_0^{-1} \leq 1$. Therefore $p_n(1) \leq p_{n-1}(1)$. Rewrite (13) in the form

$$\begin{aligned}
\gamma_n c_n &= [(\alpha_n + \gamma_n) - (\alpha_{n-1} + \gamma_{n-1})]p_n^2(1) \\
&\quad + (\alpha_n - \alpha_{n-1})[p_{n-1}^2(1) - p_n^2(1)] + \alpha_{n-1}c_{n-1}.
\end{aligned}$$

Thus $c_n > 0$.

We have shown that, in both cases (i) and (ii), we have $c_n > 0$ and hence the sequence $p_{n-1}(1)/p_n(1)$ is nondecreasing. Denote its limit by

r. Now plugging $x = 1$ into the recurrence relation for p_n, dividing both sides by $p_n(1)$ and taking the limits gives

$$1 = \frac{1}{2}[ar + (ar)^{-1}].$$

Thus $r = a^{-1}$. Let

$$\widetilde{p}_n(x) = \frac{p_n(x)}{p_n(1)}.$$

From the recurrence relation for p_n we obtain

$$x\widetilde{p}_n = \widetilde{\gamma}_n\widetilde{p}_{n+1} + \widetilde{\alpha}_n\widetilde{p}_{n-1}, \tag{14}$$

where

$$\widetilde{\alpha}_n = \frac{p_{n-1}(1)}{p_n(1)}\alpha_n, \qquad \widetilde{\gamma}_n = \frac{p_{n+1}(1)}{p_n(1)}\gamma_n.$$

By plugging $x = 1$ into (14) we get

$$\widetilde{\alpha}_n + \widetilde{\gamma}_n = 1.$$

Since both $p_{n-1}(1)/p_n(1)$ and α_n are nondecreasing, so is $\widetilde{\alpha}_n$. Moreover it tends to $1/2$ at infinity because the first of its factors tends to a^{-1} while the second tends to $a/2$. Therefore the polynomials $\widetilde{p}_n$ satisfy the assumptions of Theorem 1(i). This completes the proof. ■

4. The best normalization

Assume that the polynomials p_n satisfy (4) and (5). By multiplying each p_n by a positive constant σ_n we obtain polynomials $p_n^{(\sigma_n)}(x) = \sigma_n p_n(x)$. The positivity of Turán's determinant for the polynomials p_n is not equivalent to that for the polynomials $p_n^{(\sigma_n)}$. However, it is possible that the positivity of Turan's determinants in one normalization implies the positivity in other normalizations. It turns out that the normalization at the right most end of the interval of orthogonality has this feature.

Proposition 2. *Let the polynomials p_n satisfy (4) and (5). Assume that*

$$p_n^2(x) - p_{n-1}(x)p_{n+1}(x) \geq 0, \quad -1 \leq x \leq 1, \quad n \geq 1.$$

Let $p_n^{(\sigma)}(x) = \sigma_n p_n(x)$, where σ_n is a sequence of positive constants. Then

$$\{p_n^{(\sigma)}(x)\}^2 - p_{n-1}^{(\sigma)}(x)p_{n+1}^{(\sigma)}(x) \geq 0, \quad -1 \leq x \leq 1, \quad n \geq 1$$

if and only if

$$\sigma_n^2 - \sigma_{n-1}\sigma_{n+1} \geq 0, \quad n \geq 1.$$

Proof. We have

$$\{p_n^{(\sigma)}(x)\}^2 - p_{n-1}^{(\sigma)}(x)p_{n+1}^{(\sigma)}(x)$$
$$= (\sigma_n^2 - \sigma_{n-1}\sigma_{n+1})p_n^2(x) + \sigma_{n-1}\sigma_{n+1}(p_n^2(x) - p_{n-1}(x)p_{n+1}(x)).$$

This shows the "if" part. On the other hand, since (3) is equivalent to $p_n(1) = 1$ for $n \geq 0$, we obtain

$$\{p_n^{(\sigma_n)}(1)\}^2 - p_{n-1}^{(\sigma_n)}(1)p_{n+1}^{(\sigma_n)}(1) = \sigma_n^2 - \sigma_{n-1}\sigma_{n+1}.$$

This shows the "only if" part. ∎

Remark. Proposition 2 says that if the Turán inequality holds for the polynomials normalized at $x = 1$ then it remains true for any other normalization if and only if it holds only at the point $x = 1$, because $p_n^{(\sigma)}(1) = \sigma_n$.

5. Applications to special symmetric polynomials

We will test Theorem 1 on three classes of polynomials: ultraspherical, q – ultraspherical and symmetric Pollaczek polynomials.

The positivity of Turán's determinants for the first case is well known (see [E, p. 209]). The ultraspherical polynomials $C_n^{(\lambda)}$ are orthogonal in the interval $(-1, 1)$ with respect to the measure $(1 - x^2)^{\lambda - (1/2)}\, dx$, where $\lambda > -\frac{1}{2}$. When normalized at $x = 1$ they satisfy the recurrence relation

$$x\tilde{C}_n^{(\lambda)} = \frac{n + 2\lambda}{2n + 2\lambda}\tilde{C}_{n+1}^{(\lambda)} + \frac{n}{2n + 2\lambda}\tilde{C}_{n-1}^{(\lambda)}.$$

It can be checked easily that Theorem 1(i) or (ii) applies according to $\lambda \geq 0$ or $\lambda \leq 0$.

Let us turn to the q–ultraspherical polynomials. They have been studied by Bustoz and Ismail [BI1] but with a normalization other than the one at the right end of the interval of orthogonality. We will exhibit that our normalization is sharper in the sense that we can derive the results of [BI1] from ours. Moreover, we will have no restrictions on the parameters other than that q be positive.

In standard normalization the q–ultraspherical polynomials are denoted by $C_n(x; \beta|q)$ and they satisfy the recurrence relation

$$2xC_n(x; \beta|q) = \frac{1 - q^{n+1}}{1 - \beta q^n}C_{n+1}(x; \beta|q) + \frac{1 - \beta^2 q^{n-1}}{1 - \beta q^n}C_{n-1}(x; \beta|q). \quad (15)$$

The orthogonality measure is known explicitly (see [AI], [AW, Thm. 2.2 and Sect. 4] or [GR, Sect. 7.4]). When $|\beta|$, $|q| < 1$ it is absolutely continuous with respect to the Lebesgue measure on the interval $[-1, 1]$.

Theorem 2. *Let $0 < q < 1$ and $|\beta| < 1$. Let $\widetilde{C}_n(x;\beta|q)$ denote the q – ultraspherical polynomials normalized at $x = 1$, i.e.*

$$\widetilde{C}_n(x;\beta|q) = \frac{C_n(x;\beta|q)}{C_n(1;\beta|q)}.$$

Let

$$\Delta_n(x;\beta|q) = \widetilde{C}^2_n(x;\beta|q) - \widetilde{C}_{n-1}(x;\beta|q)\widetilde{C}_{n+1}(x;\beta|q).$$

Then

$$\Delta_n(x;\beta|q) \geq 0 \quad \text{if and only if} \quad -1 \leq x \leq 1,$$

with equality only for $x = \pm 1$.

Proof. The main obstacle in applying Theorem 1 lies in the fact that the values $C_n(1;\beta|q)$ are not given explicitly. Therefore, we cannot give explicitly the recurrence relation for $\widetilde{C}_n(x;\beta|q)$.

We will break the proof into two subcases.

(i) $0 < \beta < 1$.
Introduce the polynomials

$$p_n(x) = \beta^{n/2} \prod_{m=1}^{n} \frac{1-\beta^2 q^{m-1}}{1-q^m} C^2_n(x;\beta|q).$$

Then by (15) we obtain

$$xp_n = \gamma_n p_{n+1} + \alpha_n p_{n-1},$$

$$\alpha_n = \beta^{1/2}\frac{1-q^n}{2(1-\beta q^n)} \qquad \gamma_n = \beta^{-1/2}\frac{1-\beta^2 q^n}{2(1-\beta q^n)}.$$

Observe that

$$\alpha_n + \gamma_n = \frac{1}{2}(\beta^{1/2} + \beta^{-1/2}).$$

Moreover α_n is nondecreasing and converges to $\frac{1}{2}\beta^{1/2}$. Finally

$$\gamma_0 = \frac{1}{2}(\beta^{1/2} + \beta^{-1/2}) > 1.$$

Therefore we can apply Theorem 1(ii) with $a = \beta^{1/2}$.

(ii) $-1 < \beta \leq 0$.
Introduce the polynomials

$$p_n(x) = \prod_{m=1}^{n} \frac{1-\beta^2 q^{m-1}}{1-q^m} C^2_n(x;\beta|q).$$

Then by (15) we obtain

$$xp_n = \gamma_n p_{n+1} + \alpha_n p_{n-1},$$

$$\alpha_n = \frac{1-\beta^2 q^n}{2(1-\beta q^n)} \qquad \gamma_n = \frac{1-q^n}{2(1-\beta q^n)}.$$

Since both α_n and γ_n are increasing sequences convergent to 1 we can apply Corollary 1(i) with $a = 1$. ∎

We turn now to the symmetric Pollaczek polynomials $P_n^\lambda(x;a)$. They are orthogonal in the interval $[-1,1]$ and satisfy the recurrence relation

$$xP_n^\lambda(x;a) = \frac{n+1}{2(n+\lambda+a)} P_{n+1}^\lambda(x;a) + \frac{n+2\lambda-1}{2(n+\lambda+a)} P_{n-1}^\lambda(x;a),$$

where the parameters satisfy $a > 0,\ \lambda > 0$. We cannot compute the value $P_n^\lambda(1;a)$ in order to pass directly to normalization at $x = 1$. Instead, we consider another auxiliary normalization. Let

$$p_n(x) = \frac{n!}{(2\lambda)_n} P_n^\lambda(x;a),$$

where $(\mu)_n = \mu(\mu+1)\dots(\mu+n-1)$. Then the polynomials p_n satisfy the recurrence relation

$$xp_n = \frac{n+2\lambda}{2(n+\lambda+a)} p_{n+1} + \frac{n}{2(n+\lambda+a)} p_{n-1}.$$

Observe that the assumptions of Corollary 1 (i) or (ii) are fulfilled according to $\lambda \ge a$ or $\lambda \le a$. Therefore we have the following.

Theorem 3. *Let $\lambda > 0$, $a > 0$. Let $\widetilde{P}_n^\lambda(x;a)$ denote the Pollaczek polynomials normalized at $x = 1$, i.e.*

$$\widetilde{P}_n^\lambda(x;a) = \frac{P_n^\lambda(x;a)}{P_n^\lambda(1;a)} .$$

Then

$$\{\widetilde{P}_n^\lambda(x;a)\}^2 - \widetilde{P}_{n-1}^\lambda(x;a)\widetilde{P}_{n+1}^\lambda(x;a) \ge 0 \quad \textit{if and only if} \quad -1 \le x \le 1,$$

with equality only for $x = \pm 1$.

6. Nonsymmetric polynomials orthogonal in [-1,1]

In this section we assume that polynomials p_n satisfy (4) and (5) with β_n not necessarily equal to 0 for all n.

Theorem 4. *Let polynomials p_n satisfy (4) and (5). Let*

$$|\gamma_0 - \gamma_1| \leq \alpha_1\gamma_0 - (\gamma_0 - \gamma_1)(1 - \gamma_0). \tag{16}$$

Assume that for each $n \geq 2$ one of the following four conditions is satisfied.

(i)

$$\alpha_{n-1} \leq \alpha_n \leq \gamma_n \leq \gamma_{n-1},$$

$$\frac{\beta_n + 1 + \sqrt{(\beta_n+1)^2 - 4\alpha_n\gamma_n}}{2\gamma_n} \leq \frac{\alpha_n - \alpha_{n-1}}{\gamma_{n-1} - \gamma_n}$$

or

$$(\beta_n + 1)^2 - 4\alpha_n\gamma_n < 0 .$$

(ii)

$$\alpha_{n-1} \geq \alpha_n \geq \gamma_n \geq \gamma_{n-1},$$

$$\frac{\beta_n + 1 - \sqrt{(\beta_n+1)^2 - 4\alpha_n\gamma_n}}{2\gamma_n} \geq \frac{\alpha_{n-1} - \alpha_n}{\gamma_n - \gamma_{n-1}}$$

or

$$(\beta_n + 1)^2 - 4\alpha_n\gamma_n < 0.$$

(iii)

$$\alpha_{n-1} \geq \alpha_n \geq \frac{1}{2}, \quad \gamma_{n-1} \geq \gamma_n \geq \frac{1}{2},$$

$$\frac{\alpha_n - \alpha_{n-1}}{\gamma_n - \gamma_{n-1}} \leq \frac{\alpha_n}{\gamma_n} \leq 1 \quad \text{or} \quad \frac{\alpha_n - \alpha_{n-1}}{\gamma_n - \gamma_{n-1}} \geq \frac{\alpha_n}{\gamma_n} \geq 1.$$

(iv)

$$\alpha_{n-1} \leq \alpha_n, \quad \gamma_{n-1} \leq \gamma_n,$$

$$\begin{cases} \alpha_n \leq \gamma_n \\ \alpha_n - \alpha_{n-1} \geq \gamma_n - \gamma_{n-1} \end{cases} \quad \text{or} \quad \begin{cases} \alpha_n \geq \gamma_n \\ \alpha_n - \alpha_{n-1} \leq \gamma_n - \gamma_{n-1} \end{cases}$$

Then

$$\Delta_n(x) = p_n^2(x) - p_{n-1}(x)p_{n+1}(x) \geq 0 \quad \text{for } -1 \leq x \leq 1.$$

Proof. The proof will go by induction. Combining (8) and (4) for $n = 1$ gives

$$\gamma_0^2\gamma_1\Delta_1(x) = (1-x)[(\gamma_0-\gamma_1)(x-\beta_0)+\alpha_1\gamma_0].$$

Now using (5) gives that the positivity of $\Delta_1(x)$ in the interval $[-1,1]$ is equivalent to (16).

Fix x in $[-1,1]$ and assume that $\Delta_n(x) \geq 0$. Consider two quadratic functions

$$\begin{aligned} A(t) &= (t+1)\{(\gamma_n-\gamma_{n-1})t-(\alpha_{n-1}-\alpha_n)\}, \\ B(x;t) &= \gamma_n t^2-(\beta_n-x)t+\alpha_n. \end{aligned}$$

Set

$$t = -\frac{p_n(x)}{p_{n-1}(x)}.$$

By Proposition 1 it suffices to show that for any t the values $A(t)$ and as $B(x;t)$ cannot be both negative. In order to achieve this we have to look at the roots of these functions. The roots of $A(t)$ are -1 and $(\alpha_{n-1}-\alpha_n)/(\gamma_n-\gamma_{n-1})$; hence they are independent of x. The roots of $B(x;t)$ have always the same sign and are equal to

$$r_n^{(1)}(x) = \frac{\beta_n-x-\sqrt{(\beta_n-x)^2-4\alpha_n\gamma_n}}{2\gamma_n}, \tag{17}$$

$$r_n^{(2)}(x) = \frac{\beta_n-x+\sqrt{(\beta_n-x)^2-4\alpha_n\gamma_n}}{2\gamma_n}. \tag{18}$$

Since the function $u \mapsto u+\sqrt{u^2-a^2}$, $a>0$, is decreasing for $u \leq -a$ and increasing for $u \geq a$ we have

$$r_n^{(1)}(1) \leq r_n^{(1)}(x) \leq r_n^{(2)}(x) \leq r_n^{(2)}(1) \quad \text{if} \quad \beta_n - x \leq 0, \tag{19}$$

$$r_n^{(1)}(-1) \leq r_n^{(1)}(x) \leq r_n^{(2)}(x) \leq r_n^{(2)}(-1) \quad \text{if} \quad \beta_n - x \geq 0, \tag{20}$$

provided that $(\beta_n-x)^2-4\alpha_n\gamma_n \geq 0$. Thus $B(x;t)<0$ implies $B(1;t)<0$ ($B(-1;t)<0$ respectively) if $\beta_n-x \leq 0$ ($\beta_n-x\geq 0$ respectively). Hence it suffices to show that the values $A(t)$ and $B(1;t)$ (the values $A(t)$ and $B(-1;t)$ respectively) cannot be both negative if $\beta_n-x\leq 0$ ($\beta_n-x\geq 0$ respectively). We will break the proof into two subcases.

(a) $\beta_n - x \leq -2\sqrt{\alpha_n\gamma_n}$.
In view of (8) and (6)the roots of $B(1;t)$ are -1 and $-\frac{\alpha_n}{\gamma_n}$. By analyzing the positions of these numbers with respect to the roots of $A(t)$ one can easily verify that under each of the four assumptions (i) through (iv) the

values $A(t)$ and $B(1;t)$ cannot be both negative.

(b) $\beta_n - x \le -2\sqrt{\alpha_n\gamma_n}$.
We examine the signs of $A(t)$ and $B(-1;t)$. Consider (i), (ii) and (iii). By analysing the mutual position of the roots of $B(-1;t)$ and $A(t)$ one can verify that $A(t)$ and $B(-1;t)$ cannot be both negative.

In case (iv) we have that $B(-1;t) \ge 0$ because

$$(1+\beta_n)^2 - 4\alpha_n\gamma_n = (2-\alpha_n-\gamma_n)^2 - 4\alpha_n\gamma_n \le 0.$$

■

Remark 1. The assumption (iv) in Theorem 4 does not imply that the support of the orthogonality measure corresponding to the polynomials p_n is contained in $[-1,1]$. By (5) we have $p_n(1) = 1$ which implies that the support is located to the left of 1. However, it can extend to the left side beyond -1.

Remark 2. If we assume that $\beta_n = 0$ for $n \ge 0$, then Theorem 4 reduces to Theorem 1. Indeed, in this case we have

$$\begin{aligned} \frac{\beta_n + 1 - \sqrt{(\beta_n+1)^2 - 4\alpha_n\gamma_n}}{2\gamma_n} &= \min\left(1, \frac{\alpha_n}{\gamma_n}\right), \\ \frac{\beta_n + 1 + \sqrt{(\beta_n+1)^2 - 4\alpha_n\gamma_n}}{2\gamma_n} &= \max\left(1, \frac{\alpha_n}{\gamma_n}\right). \end{aligned}$$

Example. Set

$$\alpha_n = \frac{1}{2} - \frac{1}{n+2}, \quad \gamma_n = \frac{1}{2} + \frac{1}{2(n+2)}, \quad \beta_n = \frac{1}{2(n+2)}.$$

We can check easily that condition (16) is satisfied. We will check that also the assumptions (iii) are satisfied for every $n \ge 2$. Clearly we have $\alpha_{n-1} \le \alpha_n \le \gamma_n \le \gamma_{n-1}$. Moreover

$$\begin{aligned} r_n^{(2)}(-1) &= \frac{\beta_n + 1 + \sqrt{(1+\beta_n)^2 - 4\alpha_n\gamma_n}}{2\gamma_n} \\ &\le \frac{\beta_n + 1}{\gamma_n} \le 2 = \frac{\alpha_n - \alpha_{n-1}}{\gamma_{n-1} - \gamma_n}. \end{aligned}$$

Let $p_n(x)$ satisfy (4). By Theorem 4(iii)

$$p_n^2(x) - p_{n-1}(x)p_{n+1}(x) \ge 0 \quad \text{for} \quad -1 \le x \le 1.$$

Let us determine the interval of orthogonality. Since $\alpha_n + \beta_n + \gamma_n = 1$ we have $p_n(1) = 1$. Thus the support of the corresponding orthogonality measure is located to the left of 1. Actually the support is contained in the interval $[-1, 1]$. Indeed, it suffices to show that $c_n = (-1)^n p_n(-1) > 0$. We will show that $c_n \geq c_{n-1} > 0$ by induction. We have $c_0 = 1$. Assume $c_n \geq c_{n-1} > 0$. Then by (4)

$$\begin{aligned} \gamma_n c_{n+1} &= (1 + \beta_n) c_n - \alpha_n c_{n-1} \geq (1 + \beta_n - \alpha_n) c_n \\ &\geq (1 - \beta_n - \alpha_n) c_n = \gamma_n c_n. \end{aligned}$$

Thus $c_{n+1} \geq c_n > 0$.

7. Polynomials orthogonal in the interval $[0, +\infty)$.

Let p_n be polynomials orthogonal in the positive half axis normalized at $x = 0$, i.e. $p_n(0) = 1$. Then they satisfy the recurrence relation of the form

$$xp_n = -\gamma_n p_{n+1} + (\alpha_n + \gamma_n) p_n - \alpha_n p_{n-1}, \quad n = 0, 1, \ldots, \tag{21}$$

with initial conditions $p_{-1} = 0$, $p_0 = 1$, where α_n, and γ_n are given sequences of real coefficients such that

$$\alpha_0 = 0, \ \gamma_0 = 1, \ \alpha_{n+1} > 0, \ \gamma_n > 0, \ \text{for } n = 0, 1, \ldots. \tag{22}$$

Theorem 5. *Let polynomials p_n satisfy (21) and (22), and let*

$$\alpha_{n-1} \leq \alpha_n, \quad \gamma_{n-1} \leq \gamma_n \quad \textit{for } n \geq 1.$$

Assume that one of the following two conditions is satisfied.

(i) $\alpha_n \leq \gamma_n \qquad \alpha_n - \alpha_{n-1} \geq \gamma_n - \gamma_{n-1}.$

(ii) $\alpha_n \geq \gamma_n \qquad \alpha_n - \alpha_{n-1} \leq \gamma_n - \gamma_{n-1}.$

Then

$$\Delta_n(x) = p_n^2(x) - p_{n-1}(x) p_{n+1}(x) \geq 0 \quad \text{for } x \geq 0.$$

Proof. Let $q_n(x) = p_n(1 - x)$. Then by (21) we obtain

$$xq_n = \gamma_n q_{n+1} + (1 - \alpha_n - \gamma_n) p_n + \alpha_n q_{n-1}.$$

We have $q_n(1) = 1$. Thus the assumptions (iv) of Theorem 4 are satisfied for every n. From the proof of Theorem 4(iv) it follows that

$q_n^2(x) - q_{n-1}(x)q_{n+1}(x) \geq 0$ for $x \leq 1$ (the assumption $x \geq -1$ is inessential). Taking into account the relation between p_n and q_n gives the conclusion. ∎

A special case of Theorem 5 is when $\alpha_n - \alpha_{n-1} = \gamma_n - \gamma_{n-1}$ for every n. In this case, applying (8) gives the following.

Proposition 3. *Let polynomials p_n satisfy (21) and (22), and let*

$$\alpha_n - \alpha_{n-1} = \gamma_n - \gamma_{n-1}, \qquad n \geq 1.$$

Then

$$p_n^2(x) - p_{n-1}(x)p_{n+1}(x) = \sum_{k=1}^{n} \frac{(\alpha_k - \alpha_{k-1})\alpha_k\alpha_{k+1}\ldots\alpha_{n-1}}{\gamma_k\gamma_{k+1}\cdots\gamma_n}(p_k(x) - p_{k-1}(x))^2.$$

In particular, if $\alpha_n \geq \alpha_{n-1}$ for $n \geq 1$, then

$$p_n^2(x) - p_{n-1}(x)p_{n+1}(x) \geq 0 \quad \textit{for} \ -\infty < x < \infty,$$

where equality holds only for $x = 0$.

Example.

Let $p_n(x) = L_n^\alpha(x)/L_n^\alpha(1)$, where $L_n^\alpha(x)$ denote the Laguerre polynomials of order $\alpha > -1$. Then the polynomials p_n satisfy

$$xp_n = -(n + \alpha + 1)p_{n+1} + (2n + \alpha + 1)p_n - np_n.$$

Then

$$\alpha_n - \alpha_{n-1} = \gamma_n - \gamma_{n-1} = 1, \quad n \geq 1.$$

Thus Proposition 3 applies. The formula for $p_n^2 - p_{n-1}p_{n+1}$ in this case is not new. It has been discovered by V. R. Thiruvenkatachar and T. S. Nanjundiah [TN] (see also [AC, 4.7]).

Acknowledgements. I am grateful to J. Bustoz and M. E. H. Ismail for kindly sending me a preprint of [BI2]. I thank George Gasper for pointing out the references [AC, TN].

References

[AC] Al-Salam W.; Carlitz, L.: General Turán expressions for certain hypergeometric series, Portugal. Math. **16**, 119–127 (1957).

[A] Askey, R.: Linearization of the product of orthogonal polynomials, Problems in Analysis, R. Gunning, ed., Princeton University Press, Princeton, N.J., 223–228 (1970).

[AI] Askey, R.,Ismail, M.E.H.: A generalization of ultraspherical polynomials, in Studies in Pure Mathematics, P. Erdös, ed., Birkhäuser, Basel, 55–78 (1983).

[AW] Askey, R., Wilson, J.A.: Some basic hypergeometric orthogonal polynomials that generalize Jacobi polynomials, Mem. Amer. Math. Soc. **54** (1985).

[BI1] Bustoz, J., Ismail, M.E.H.: Turaán inequalities for ultraspherical and continuous q–ultraspherical polynomials, SIAM J. Math. Anal. **14**, 807–818 (1983).

[BI2] Bustoz, J., Ismail, M.E.H.: Turán inequalities for symmetric orthogonal polynomials, preprint (1995).

[E] Erdélyi, A.: "Higher transcendental functions," vol. 2, New York, 1953.

[G1] Gasper, G.: On the extension of Turán's inequality to Jacobi polynomials, Duke. Math. J. **38**, 415–428 (1971).

[G2] Gasper, G.: An inequality of Turán type for Jacobi polynomials, Proc. Amer. Math. Soc. **32**, 435–439 (1972).

[GR] Gasper, G., Rahman, M.: "Basic Hypergeometric Series," Vol. 35, Encyclopedia of Mathematics and Its Applications, Cambridge University Press, Cambridge, 1990.

[KS] Karlin, S.; Szegö, G.: On certain determinants whose elements are orthogonal polynomials, J. d'Analyse Math. **8**, 1–157 (1960/61).

[MN] Mukherjee, B.N; Nanjundiah, T.S.: On an inequality relating to Laguerre and Hermite polynomials, Math. Student. **19**, 47–48 (1951).

[Sk] Skovgaard, H.: On inequalities of the Turán type, Math. Scand. **2**, 65–73 (1954).

[S] Szász, O.: Identities and inequalities concerning orthogonal polynomials and Bessel functions, J. d'Analyse Math. **1**, 116–134 (1951).

[Sz1] Szegö, G.: On an inequality of P. Turán concerning Legendre polynomials, Bull. Amer. Math. Soc. **54**, 401–405 (1948).

[Sz2] Szegö, G.: An inequality for Jacobi polynomials, Studies in Math. Anal. and Related Topics, Stanford Univ. Press, Stanford, Calif., 392–398 (1962).

[TN] Thiruvenkatachar, V.R.; Nanjundiah, T.S.: Inequalities concerning Bessel functions and orthogonal polynomials, Proc. Indian Acad. Sci. **33**, 373–384 (1951).

[T] Turán, P.: On the zeros of the polynomials of Legendre, Časopis Pěst. Mat. **75**, 113–122 (1950).

Institute of Mathematics
Polish Academy of Sciences
ul. Kopernika
00–950 Wrocław, Poland

Current address:
Institute of Mathematics
Wrocław University
pl. Grunwaldzki 2/4
50–384 Wrocław, Poland

Wavelets on Hypergroups

K. Trimeche

Abstract

We consider hypergroups K satisfying certain conditions. Important examples of such hypergroups are the double coset hypergroup, the Chébli–Trimèche hypergroup and the hypergroup associated with spherical mean operator. We define on K wavelets and a continuous wavelet transform, we prove Plancherel and inversion formulas for this transform, and using coherent states we characterize the image space of this transform.

Introduction

Hypergroups are locally compact spaces with a group-like structure on which the bounded measures convolve in a similar way to that on a locally compact group. In fact, a hypergroup K can be viewed as a probabilistic group in the sense that to each pair x, y of points of K there exists a probability measure $\delta_x * \delta_y$ on K with compact support, not necessarily equal to the Dirac measure δ_{xy}, for a composition x, y in K, such that $(x, y) \to \mathrm{Supp}\,(\delta_x * \delta_y)$ is a continuous mapping from $K \times K$ into the space of compact subsets of K. In place of the natural left translation of a function f by x, available in the group case, we deal in a hypergroup with a generalized (left) translation defined by

$$(1) \qquad T_x(f)(y) = \int_K f(z) d(\delta_x * \delta_y)(z)$$

for all y in K.

For commutative hypergroups K a substantial body of Harmonic Analysis has been built, where the convolution product, Fourier transform and Fourier–Plancherel transform are available as important technical tools (see [1], [8]). In this work we consider commutative hypergroups for which their Plancherel measure satisfies certain conditions. Important examples of such hypergroups are

- The double coset hypergroup $K = \mathcal{K}\backslash G/\mathcal{K}$ where G is a noncompact connected real semisimple Lie group with finite center and $\mathcal{K}$ is a compact subgroup of G (see [1], [6], [7]).
- The space $K = [0, +\infty[$ with operation inherited from a singular differential operator. Such a hypergroup is called a Chébli–Trimèche hypergroup (see [1], [2], [18], [19], [20]).
- The space $K = [0, +\infty[\times\mathbb{R}^n$ with operation inherited from the spherical mean operator (see [11], [15]).

The continuous wavelet transform for functions on $\mathbb{R}^n$ is an integral transform for which the kernel is the dilated translate of a so-called wavelet g, a quite arbitrary square integrable function on $\mathbb{R}^n$. A Plancherel formula for this transform is obtained, and in the course of its derivation we naturally arrive at an admissibility assumption for the wavelet g. There follow a Parseval and an inversion formula. Wavelet transforms have widespread applications, ranging from signal analysis in geophysics and acoustics, to quantum theory and pure mathematics. (See [3], [9], [10].)

It is a natural question to ask whether there exist continuous wavelet transforms on hypergroups.

Using the generalized translation (1) and the dilation obtained by Fourier transform we show that it is possible to define continuous wavelet transforms on hypergroups. We prove for these transforms Plancherel and inversion formulas, and we discuss them in the framework of coherent states.

We conclude this introduction with a summary of the content of this paper.

In the first section we recall some results about hypergroups and the Harmonic Analysis on them (convolution product, Fourier transform, Plancherel and inversion formulas, Plancherel theorem, ...).

In the second section we give the three examples of hypergroups cited above.

In the third section we define wavelets and continuous wavelet transforms on hypergroups and we prove for these transforms Plancherel and inversion formulas, and using coherent states we characterize the image spaces of these transforms.

1. Harmonic analysis on hypergroups

1.1. Definition of hypergroups

Let K be a locally compact topological space with an operation $*$, called convolution, on the space $M(K)$ of bounded complex Radon measures on K, which is a bilinear and separately continuous mapping from $M(K) \times M(K)$ into $M(K)$ which preserves probability measures (i.e. $M_1(K) \times M_1(K) \subset M_1(K)$). $(K, *)$ is a hypergroup in the sense of Jewett (see [8]) if the following properties are satisfied for all $x, y, z \in K$:

C_1: $\delta_x * (\delta_y * \delta_z) = (\delta_x * \delta_y) * \delta_z$, where δ_x is the Dirac measure at the point x.

C_2: There exists an element $e \in K$ such that

$$\delta_e * \delta_x = \delta_x * \delta_e = \delta_x.$$

C_3: There exists a continuous involution $x \to x^-$ of K such that $e \in \mathrm{Supp}\,(\delta_x * \delta_y)$ if and only if $x = y^-$ and the canonical extension of this involution to the space $M(K)$ satisfies

$$(\delta_x * \delta_y)^- = \delta_{y^-} * \delta_{x^-}.$$

C_4: The mapping $(x, y) \to \delta_x * \delta_y$ is weakly continuous from $K \times K$ into $M(K)$.

C_5: The support of $\delta_x * \delta_y$ is compact.

C_6: The mapping $(x, y) \to \mathrm{Supp}\,(\delta_x * \delta_y)$ from $K \times K$ into the set of non-empty compact subsets of K is continuous.

The hypergroup $(K, *)$ is commutative if

$$\delta_x * \delta_y = \delta_y * \delta_x, \quad \text{for all} \quad x, y \in K.$$

A particular case is when $x = x^-$, and we say then that K is hermitian.

In the following sections we consider only commutative hypergroups K. A commutative hypergroup always has a Haar measure, i.e. a positive Radon measure m such that for all f continuous on K and with compact support we have

$$\forall\, x \in K, \quad \langle m, \delta_x * f \rangle = \langle m, f \rangle$$

where $\delta_x * f$ denotes the function defined by

$$(\delta_x * f)(y) = \langle \delta_y * \delta_{x^-}, f \rangle.$$

The measure m is unique up to a multiplicative constant and $\operatorname{Supp} m = K$ (see [1], [8]).

1.2. Generalized translation operator and convolution on hypergroups

Notations: We denote by

(i) $C(K)$ the space of continuous functions on K.

(ii) $C_c(K)$ the subspace of $C(K)$ consisting of compactly supported functions.

(iii) $L^p(K, m)$, $p \in [1, +\infty]$, the space of measurable functions on K such that

$$\|f\|_p = \left(\int_K |f(x)|^p dm(x) \right)^{1/p} < +\infty, \; 1 \leq p < +\infty,$$
$$\|f\|_\infty = \operatorname*{ess\,sup}_{x \in K} |f(x)| < +\infty.$$

Definition 1.1. Let f be in $C(K)$ and $x, y \in K$. We define the generalized translation operator T_x by

$$T_x(f)(y) = \int_K f(z) d(\delta_x * \delta_y)(z).$$

Properties.

We have the following properties:

(i) $T_x(f)(y) = T_y(f)(x)$, for all $f \in C(K)$ and $x, y \in K$.

(ii) If f is in $C_c(K)$, then for all $x \in K$, the mapping $y \to T_x(f)(y)$ belongs to $C_c(K)$.

(iii) $\forall \; x, y \in K$, $T_x(1)(y) = 1$.

(iv) $\forall \; x \in K$, $\forall \; f \in L^p(K, m)$, $p \in [1, +\infty]$, $\|T_x(f)\|_p \leq \|f\|_p$.

(v) For f in $L^p(K, m)$, $p \in [1, +\infty[$, the mapping $x \to T_x(f)$ is continuous from K into $L^p(K, m)$, $p \in [1, +\infty[$.

Definition 1.2.

(i) Let μ and ν be in $M(K)$. The convolution of μ and ν is the measure $\mu * \nu$ in $M(K)$ defined by

$$\mu * \nu(f) = \int_K T_x(f)(y) d\mu(x) d\nu(y), \quad f \in C_c(K).$$

(ii) If $\mu = gm$ and $\nu = hm$ with g, h in $L^1(K, m)$, then

$$\mu * \nu = (gm) * (hm) = (g * h)m$$

where $g * h$ is the convolution of g and h given by

$$g * h(x) = \int_K T_x(g)(y)h(y^-)dm(y).$$

Proposition 1.1. *If g is in $L^p(K, m)$ and h is in $L^q(K, m)$ with $p, q \in [1, +\infty]$, then the function $g*h$ belongs to $L^r(K, m)$, $r \in [1, +\infty]$, with $\frac{1}{p} + \frac{1}{q} - 1 = \frac{1}{r}$ and we have*

$$\|g * h\|_r \leq \|g\|_p \|h\|_q$$

(see [8]).

1.3. Fourier transform on hypergroups.

A complex valued function χ on K will be called a multiplicative function if χ is continuous, not identically zero, and has the property that

$$\forall\, x, y \in K,\ T_x(\chi)(y) = \chi(x)\chi(y);$$

we say that χ is hermitian if

$$\forall\, x \in K,\ \chi(x^-) = \overline{\chi(x)}.$$

The dual of a commutative hypergroup $(K, *)$ is the space $\widehat{K}$ of all bounded multiplicative and hermitian functions χ on K, endowed with the compact open topology. Note that $\widehat{K}$ is nonvoid, since it contains the constant function 1, and we have

$$\sup_{x \in K} |\chi(x)| = \chi(e) = 1.$$

Definition 1.3. (i) For μ in $M(K)$, the Fourier transform $\hat{\mu}$ of μ is defined on $\widehat{K}$ by

$$\hat{\mu}(\chi) = \int_K \overline{\chi(x)} d\mu(x).$$

(ii) For f in $L^1(K,m)$, we have

$$(fm)^\wedge(\chi) = \int_K f(x)\overline{\chi(x)}dm(x);$$

we denote the first member $\hat{f}(\chi)$ and we call $\hat{f}$ the Fourier transform of f.

Properties.

(i) For μ in $M(K)$ the function $\hat{\mu}$ is continuous on $\widehat{K}$ and we have

$$\forall\, \chi \in \widehat{K},\ |\hat{\mu}(\chi)| \leq \|\mu\|$$

where $\|\mu\| = \int_K d|\mu|$ is the norm of the measure μ.

(ii) For μ and ν in $M(K)$ we have

$$\forall\, \chi \in \widehat{K},\ (\mu * \nu)^\wedge(\chi) = \hat{\mu}(\chi)\hat{\nu}(\chi).$$

(iii) For f in $L^1(K,m)$ the function $\hat{f}$ is continuous on $\widehat{K}$, zero at infinity and we have

$$\forall\, \chi \in \widehat{K},\ |\hat{f}(\chi)| \leq \|f\|_1.$$

(iv) For f and g in $L^1(K,m)$ we have

$$\forall\, \chi \in \widehat{K},\ (f * g)^\wedge(\chi) = \hat{f}(\chi)\hat{g}(\chi)$$

(see [8]).

Theorem 1.1. *There exists a unique positive Radon measure π on $\widehat{K}$, called the Plancherel measure of K, such that for every f and g in $(L^1 \cap L^2)(K,m)$*

$$\int_K f(x)\overline{g(x)}dm(x) = \int_{\widehat{K}} \hat{f}(\chi)\overline{\hat{g}(\chi)}d\pi(\chi).$$

Notations: We denote by

- $L^p(\widehat{K}, \pi)$, $p \in [1, +\infty]$, the space of measurable functions on $\widehat{K}$ such that

$$\|f\|_{p,\pi} = \left(\int_{\widehat{K}} |f(\chi)|^p d\pi(\chi) \right)^{1/p}, \quad 1 \leq p < +\infty$$

$$\text{and} \quad \|f\|_{\infty,\pi} = \operatorname*{ess\,sup}_{\chi \in \widehat{K}} |f(\chi)| < +\infty.$$

- $\mathcal{S}(K)$ (resp. $\mathcal{S}(\widehat{K})$) the space of simple functions on K (resp. $\widehat{K}$) which vanish outside a set of finite measure.

Theorem 1.2. *The mapping $f \to \hat{f}$ from $(L^1 \cap L^2)(K, m)$ into $L^2(\widehat{K}, \pi)$ extends to an isometric isomorphism from $L^2(K, m)$ onto $L^2(\widehat{K}, \pi)$ (see [8], p. 41).*

Corollary 1.1.
(i) Let f be in $L^1(K, m)$. Then for all $x \in K$ we have

$$(T_x(f))^\wedge(\chi) = \chi(x)\hat{f}(\chi).$$

(ii) Let f be in $L^1(K, m)$ and g in $L^2(K, m)$. Then

$$(f * g)^\wedge = \hat{f} \cdot \hat{g}.$$

Theorem 1.3. *Let f be in $C(K) \cap L^1(K, m)$ such that f belongs to $L^1(\widehat{K}, \pi)$. Then we have the inversion formula*

$$f(x) = \int_{\widehat{K}} \hat{f}(\chi)\chi(x) d\pi(\chi), \quad x \in K$$

(see [8], p. 73).

Proposition 1.2. *For all f and g in $L^2(K, m)$ and for all Ψ in $\mathcal{S}(K)$ and θ in $\mathcal{S}(\widehat{K})$ we have the relations*

$$\int_K f * g(x)\Psi(x) dm(x) = \int_{\widehat{K}} \hat{f}(\chi)\hat{g}(\chi)\widehat{\Psi}(\chi) d\pi(\chi) \tag{1.1}$$

$$\int_K f * g(x)\widehat{\theta}(x) dm(x) = \int_{\widehat{K}} \hat{f}(\chi)\hat{g}(\chi)\theta(\chi) d\pi(\chi). \tag{1.2}$$

Proof. Let $g \in L^2(K,m)$ and $\Psi \in \mathcal{S}(K)$. We consider the operators ℓ_1 and ℓ_2 defined on $L^2(K,m)$ by

$$\ell_1(f) = \int_K f * g(x)\Psi(x)dm(x)$$
$$\ell_2(f) = \int_{\widehat{K}} \hat{f}(\chi)\hat{g}(\chi)\widehat{\Psi}(\chi)d\pi(\chi).$$

From Proposition 1.1, Corollary 1.1 and Theorem 1.1 we deduce that for all $f \in (L^1 \cap L^2)(K,m)$ we have

$$\ell_1(f) = \ell_2(f).$$

But $(L^1 \cap L^2)(K,m)$ is dense in $L^2(K,m)$. Then to prove that $\ell_1 \equiv \ell_2$ it suffices to show that ℓ_1 and ℓ_2 are continuous on $L^2(K,m)$.

Using Proposition 1.1, Hölder's inequality and Theorem 1.1 we obtain

$$\begin{aligned}|\ell_1(f)| &\le \|f * g\|_\infty \|\Psi\|_1\\ &\le \|f\|_2\|g\|_2\|\Psi\|_1\end{aligned}$$

and

$$\begin{aligned}|\ell_2(f)| &\le \|\hat{f}\cdot\hat{g}\|_{1,\pi}\|\widehat{\Psi}\|_{\infty,\pi}\\ &\le \|\hat{f}\|_{2,\pi}\|\hat{g}\|_{2,\pi}\|\widehat{\Psi}\|_{\infty,\pi}\\ &= \|f\|_2\|g\|_2\|\widehat{\Psi}\|_{\infty,\pi}.\end{aligned}$$

This completes the proof of the relation (1.1).

We obtain the relation (1.2) by the same proof as for the relation (1.1). ■

Theorem 1.4. *Let f and g be in $L^2(K,m)$. Then the function $f * g$ belongs to $L^2(K,m)$ if and only if the function $\hat{f}\cdot\hat{g}$ belongs to $L^2(\widehat{K},\pi)$ and we have*

$$(1.3) \qquad (f * g)^\wedge = \hat{f}\cdot\hat{g}.$$

Proof. (i) We suppose that $f * g \in L^2(K,m)$. As f and g are in $L^2(K,m)$, from Theorem 1.2 the function $\hat{f}\cdot\hat{g}$ belongs to $L^1(\widehat{K},\pi)$.

Thus for all $\theta \in \mathcal{S}(\widehat{K})$ the function $\hat{f} \cdot \hat{g} \cdot \theta$ belongs also to $L^1(\widehat{K}, \pi)$. From the relation (1.2), Hölder's inequality and Theorem 1.2 we deduce

$$\left| \int_{\widehat{K}} \hat{f}(\chi)\hat{g}(\chi)\theta(\chi)d\pi(\chi) \right| \leqq \|f * g\|_2 \|\check{\theta}\|_2 = \|f * g\|_2 \|\theta\|_{2,\pi} \ ;$$

thus the quantity

$$\operatorname{Sup}\left\{ \left| \int_{\widehat{K}} \hat{f}(\chi)\hat{g}(\chi)\theta(\chi)d\pi(\chi) \right|, \theta \in \mathcal{S}(\widehat{K}), \|\theta\|_{2,\pi} = 1 \right\}$$

is finite. From the converse of Hölder's inequality (see [5], p. 181) we deduce that $\hat{f} \cdot \hat{g} \in L^2(\widehat{K}, \pi)$. We obtain the relation (1.3) from the relation (1.2), Theorem 1.2 and the density of $\mathcal{S}(\widehat{K})$ in $L^2(\widehat{K}, \pi)$.

(ii) We suppose that $\hat{f} \cdot \hat{g} \in L^2(\widehat{K}, \pi)$. As f and g are in $L^2(K, m)$ then, from Proposition 1.1, the function $f * g$ belongs to $L^\infty(K, m)$; thus for all $\Psi \in \mathcal{S}(K)$ the function $f * g \cdot \Psi$ belongs to $L^1(K, m)$. From the relation (1.1), Hölder's inequality and Theorem 1.2 we deduce

$$\begin{aligned} \left| \int_K f * g(x)\Psi(x)dm(x) \right| &\leq \|\hat{f} \cdot \hat{g}\|_{2,\pi} \|\widehat{\Psi}\|_{2,\pi} \\ &= \|\hat{f} \cdot \hat{g}\|_{2,\pi} \|\Psi\|_2; \end{aligned}$$

we deduce the results by the same proof as for (i). ∎

Corollary 1.2. *For f and g in $L^2(K, m)$ we have*

$$\int_K |f * g(x)|^2 dm(x) = \int_{\widehat{K}} |\hat{f}(\chi)|^2 |\hat{g}(\chi)|^2 d\pi(\chi); \tag{1.4}$$

both sides are finite or infinite.

Proof. For $f * g$ in $L^2(K, m)$ the relation (1.4) can be deduced from Theorem 1.2. For the other case the two of sides (1.4) are infinite. ∎

In this work we suppose that the commutative hypergroup K satisfies the following conditions

(1.5) (i) The dual $\widehat{K}$ can be identified with a subset Γ of $\mathbb{C}^n$, $n \geq 1$, which contains $\mathbb{R}^n$. We denote by φ_λ, $\lambda \in \Gamma$, the elements of $\widehat{K}$; in particular we denote φ_{λ_0} the element corresponding to the character 1. We suppose that for all $x \in K$, the mapping $\lambda \to \varphi_\lambda(x)$ is continuous on Γ.

(1.6) (ii) The Plancherel measure $d\pi$ is absolutely continuous with respect to the Lebesgue measure of $\mathbb{R}^n$, with density π; we denote its support by S. We suppose that S and π satisfy

- For all λ in S the function $\varphi_\lambda(x)$ is real.
- π is continuous on S and there exist $\gamma \in \mathbb{R}$ such that

$$\forall\, \lambda \in S,\ |\pi(\lambda)| \leq \text{const. } (1 + \|\lambda\|)^\gamma$$

where

$$\|\lambda\|^2 = \sum_{j=1}^{n} |\lambda_j|^2, \quad \text{if} \quad \lambda = (\lambda_1, \ldots, \lambda_n) \in S.$$

- S contain the point 0, and for all $a > 0$, we have: $aS = S$.
- $\forall\, \lambda \in S \backslash \{0\}$, $\pi(\lambda) > 0$.
- The function k defined on $]0, +\infty[$ by

$$k(a) = \sup_{\lambda \in S\backslash\{0\}} \frac{\pi\left(\frac{\lambda}{a}\right)}{\pi(\lambda)}$$

is continuous on $]0, +\infty[$. ■

Notations. Using conditions (1.5) and (1.6) we denote by

- $\mathcal{F}(\mu)(\lambda)$ and $\mathcal{F}(f)(\lambda)$ respectively the Fourier transform $\hat{\mu}$ and $\hat{f}$ of μ and f.
- $L^p(S, \pi)$, $p \in [1, +\infty]$, the space $L^p(\widehat{K}, \pi)$, $p \in [1, +\infty]$.

2. Examples of hypergroups.

In this section we shall give three examples of hypergroups satisfying the conditions (1.5) and (1.6) of the previous section.

2.1. The double coset hypergroup $K = \mathcal{K}\backslash G/\mathcal{K}$

Let G be a noncompact connected real semisimple Lie group with finite center, and let $\mathcal{G}$ be the Lie algebra of G. Let $\mathcal{G} = k + p$ be the Cartan decomposition of $\mathcal{G}$. Let $\mathcal{A} \subset p$ be a maximal abelian subspace, $\mathcal{A}^*$ its (real) dual, $\mathcal{A}^*_{\mathbb{C}}$ the complexification of $\mathcal{A}^*$. The dimension n of $\mathcal{A}$ is called the real rank of G. We can identify $\mathcal{A}$ with $\mathbb{R}^n$. We denote by $\|\cdot\|$ the norm in $\mathcal{A}$ and by Σ the set of restricted roots. Let W be the Weyl group associated with Σ and $|W|$ its cardinality.

Fix a Weyl chamber $\mathcal{A}^+$ in $\mathcal{A}$ and let $\overline{\mathcal{A}}^+$ be its closure. We call a root positive if it is positive on $\mathcal{A}^+$. Let Σ^+ be the set of positive roots, and put $\Sigma_0^+ = \Sigma^+ \cap \left\{\alpha \in \Sigma \middle/ \frac{1}{2}\,\alpha \in \Sigma\right\}$. Put $\rho = \frac{1}{2} \sum\limits_{\alpha\in\Sigma^+} m_\alpha \cdot \alpha$, with m_α the multiplicity of α. Let $\mathcal{K}$ (resp. A) be the analytic subgroup of G with Lie algebra k (resp. $\mathcal{A}$), and $H : G \to \mathcal{A}$ the Iwasawa projection according to the Iwasawa decomposition $G = \mathcal{K}AN$, i.e. if $g \in G$, then $H(g)$ is the unique element in $\mathcal{A}$ such that $g \in \mathcal{K}(\exp H(g))N$. On the compact group $\mathcal{K}$ the Haar measure dk is normalized by $\int_{\mathcal{K}} dk = 1$. In the Cartan decomposition $G = \mathcal{K} \exp \overline{\mathcal{A}}^+ \mathcal{K}$, and the Haar measure dx of G is given by

$$\int_G f(x)dx = \int_{\mathcal{K}} \int_{\mathcal{A}^+} \int_{\mathcal{K}} f(k_1(\exp H)k_2)\delta(H)dk_1 dH dk_2$$

where f is a continuous function on G with compact support and $\delta(H) = \prod_{\alpha\in\Sigma^+}[2\mathrm{sh}\alpha(H)]^{m_\alpha}$; here and throughout sh and ch represent the hyperbolic sine and hyberbolic cosine. We have the following estimate for the density $\delta(H)$:

$$0 \le \delta(H) \le e^{2\rho(H)}, \quad (H \in \overline{\mathcal{A}}^+).$$

The spherical functions on G are the functions

$$\varphi_\lambda(x) = \int_{\mathcal{K}} e^{(i\lambda-\rho)(H(xk))} dk, \quad x \in G,\ \lambda \in \mathcal{A}_{\mathbb{C}}^*.$$

They satisfy the following properties

(i) The function $\varphi_\lambda(x)$ is a C^∞-function in x and a holomorphic function in λ.

(ii) We have

- $\forall\ \lambda \in \mathcal{A}_{\mathbb{C}}^*$, $\varphi_\lambda(e) = 1$
- $\forall\ x \in G$, $\varphi_{i\rho}(x) = 1$.

(iii) We have the product formula

$$\forall \quad x, y \in G,\ \varphi_\lambda(x)\varphi_\lambda(y) = \int_{\mathcal{K}} \varphi_\lambda(xky)dk.$$

(iv) We have

$$|\varphi_\lambda(x)| \leq \varphi_0(x), \ (x \in G, \lambda \in \mathcal{A}^*)$$

with

$$0 < \varphi_0(\exp H) \leq \text{const.}\,(1 + \|H\|)^a e^{-\rho(H)}, \ (H \in \overline{\mathcal{A}}^+)$$

for some positive constant a. (See [6], [7].)

Definition 2.1. (i) For all $x, y \in G$ and f a continuous function on G bi-invariant under $\mathcal{K}$, we put

$$\delta_{\mathcal{K}x\mathcal{K}} * \delta_{\mathcal{K}y\mathcal{K}}(f) = \int_{\mathcal{K}} f(xky)dk.$$

The mapping $*$ is called a generalized convolution product on the double coset space $K = \mathcal{K}\backslash G/\mathcal{K}$.

(ii) The involution on K is defined by

$$\forall\, x \in G, \ (\mathcal{K}x\mathcal{K})^- = \mathcal{K}x^{-1}\mathcal{K}.$$

Theorem 2.1. *(i)* $(K, *)$ *is a hypergroup in the sense of Jewett and is called a double coset hypergroup.*

(ii) The Haar measure on K *is given by the relation (2.1).*

(iii) The dual space can be identified with $\mathcal{A}^* + iC^\rho$*, where* C^ρ *is the convex hull of* ρ *defined by*

$$C^\rho = \{w \cdot \rho / w \in W\}.$$

(iv) The Plancherel measure is given by

$$d\pi(\lambda) = \frac{I}{|W|}|c(\lambda)|^{-2}d\lambda$$

where c *is the Harish–Chandra function.*

(v) We have

$$S = \operatorname{Supp}\pi = \mathcal{A}^*.$$

(vi) We have

$$k(a) = \sup_{\lambda \in \mathcal{A}^* \setminus \{0\}} \frac{|c(\lambda/a)|^{-2}}{|c(\lambda)|^{-2}} = \begin{cases} a^{-2\mathrm{card}\Sigma_0^+}, & \textit{if } a \geq 1 \\ a^{-\dim N}, & \textit{if } 0 < a < 1. \end{cases}$$

In particular, when G has a complex structure, we have

$$k(a) = a^{-2\mathrm{card}\Sigma_0^+}, \quad \textit{for all} \quad a > 0.$$

(For (i),...,(v) see [1], [6], [7], and for (vi) see [16].)

2.2. The Chébli–Trimèche hypergroups.

We consider the function A defined on $[0, +\infty[$ satisfying the following conditions

(i) $A(x) = x^{2\alpha+1}B(x)$, where $\alpha > -\frac{1}{2}$ and B is a positive even C^∞-function on $\mathbb{R}$ such that $B(0) = 1$.

(ii) A is increasing and unbounded.

(iii) $\frac{A'}{A}$ is decreasing on $[0, +\infty[$ and $\lim_{x\to+\infty} \dfrac{A'(x)}{A(x)} = 2\rho \geq 0$.

(iv) There exists a positive constant γ_0 such that for all $r \in [a, +\infty[$, $a > 0$, we have

- If $\rho > 0$: $\dfrac{B'(x)}{B(x)} = 2\rho - \dfrac{2\alpha+1}{x} + e^{-\gamma_0 x}F(x)$
- If $\rho = 0$: $\dfrac{B'(x)}{B(x)} = e^{-\gamma_0 x}F(x)$

where F is a C^∞-function on $[a, +\infty[$ bounded together with its derivatives.

Let L_A be the differential operator on $]0, +\infty[$:

$$L_A = \frac{d^2}{dx^2} + \frac{A'(x)}{A(x)} \frac{d}{dx}.$$

Examples.

(i) If

$$A(x) = x^k, \ k \in \mathbb{N}, \ k \geq 2,$$

then L_A is the radial part of the Laplacian operator on $\mathbb{R}^k$.

(ii) If

$$A(x) = (\operatorname{sh} x)^k, \ k \in \mathbb{N}, \ k \geq 2,$$

then L_A is the radial part of the Laplace–Beltrami operator on the k-dimensional hyperbolic space.

We consider the equation

$$\text{(2.2)} \qquad \begin{cases} L_A u = -(\lambda^2 + \rho^2)u, & \lambda \in \mathbb{C} \\ u(0) = 1, \quad u'(0) = 0. \end{cases}$$

Proposition 2.1. *(i) The equation (2.2) admits a unique C^∞-solution φ, which can be extended to an even function on $\mathbb{R}$.*

(ii) For each $x \in [0, +\infty[$, the function $\lambda \to \varphi_\lambda(x)$ is an entire function on $\mathbb{C}$.

(iii) For each $x \in [0, +\infty[$, we have

$$|\varphi_\lambda(x)| \leq 1$$

if and only if λ belongs to the closure of the set $\Sigma = \{\lambda \in \mathbb{C}/|Im\,\lambda| < \rho\}$.

(iv) We have

$$\forall\, x \in [0, +\infty[, \quad \varphi_{i\rho}(x) = 1.$$

(See [13], [14], [18].)

Examples.

(i) If

$$A(x) = x^{2\alpha+1}, \quad \alpha > -\frac{1}{2}$$

we have

$$\varphi_\lambda(x) = j_\alpha(\lambda x) = \begin{cases} 2^\alpha \Gamma(\alpha+1) \dfrac{J_\alpha(\lambda x)}{(\lambda x)^\alpha}, & \text{if} \quad \lambda x \neq 0 \\ 1, & \text{if} \quad \lambda x = 0 \end{cases}$$

where J_α is the Bessel function of first kind and order α.

(ii) If

$$A(x) = 2^{2(\alpha+\beta+1)} (\operatorname{sh} x)^{2\alpha+1} (\operatorname{ch} x)^{2\beta+1}, \quad \alpha \geq \beta > -\frac{1}{2}$$

we have

$$\varphi_\lambda(x) = \varphi_\lambda^{(\alpha,\beta)}(x)$$

where $\varphi_\lambda^{(\alpha,\beta)}$ is the Jacobi function defined by

$$\varphi_\lambda^{(\alpha,\beta)}(x) = {}_2F_1\left(\frac{1}{2}\,(\alpha+\beta+1-i\lambda), \frac{1}{2}\,(\alpha+\beta+1+i\lambda); -\mathrm{sh}^2 x\right)$$

with ${}_2F_1$ the Gauss hypergeometric function. (See [13], [14].)

Proposition 2.2. *The function φ_λ, $\lambda \in \mathbb{C}$, satisfies the following product formula*

$$\forall\ x, y \in]0, +\infty[, \quad \varphi_\lambda(x)\varphi_\lambda(y) = \int_{|x-y|}^{x+y} \varphi_\lambda(z) W(x,y,z) A(z) dz$$

where $W(x,y,z)$ is a nonnegative continuous function on $]|x-y|, x+y[$ supported in $[|x-y|, x+y]$, such that

$$\int_0^\infty W(x,y,z)A(z)dz = 1.$$

(See [14], [18], [19], [20].)

Examples.

(i) In the case of the normalized Bessel function j_α we have

$$W(x,y,z) = \begin{cases} \dfrac{a_\alpha[(x+y)^2 - z^2]^{\alpha-1/2}[z^2-(x-y)^2]^{\alpha-1/2}}{(xyz)^\alpha}, & \text{if } |x-y| < z < x+y \\ 0, & \text{otherwise} \end{cases}$$

where

$$a_\alpha = \frac{2^{1-2\alpha}\Gamma(\alpha+1)}{\sqrt{\pi}\,\Gamma(\alpha+1/2)}.$$

(See [13], p. 93.)

(ii) In the case of the Jacobi function $\varphi_\lambda^{(\alpha,\beta)}$ the function $W(x,y,z)$ is given by

• For $\alpha > \beta > -\frac{1}{2}$, we have

$$W(x,y,z) = \begin{cases} b_{\alpha,\beta} \dfrac{1}{(\text{sh}x\text{sh}y\text{sh}z)^{2\alpha}} \displaystyle\int_0^\pi [1 - \text{ch}^2 x - \text{ch}^2 y - \text{ch}^2 z \\ \qquad + 2\text{ch}x\text{ch}y\text{ch}z\cos\theta]_+^{\alpha-\beta-1} \\ \qquad + (\text{sh}\,\theta)^{2\beta} d\theta\ , & \text{if } |x-y| < z < x+y \\ 0\ , & \text{otherwise} \end{cases}$$

where

$$b_{\alpha,\beta} = \frac{2^{1-2(\alpha+\beta+1)}\Gamma(\alpha+1)}{\sqrt{\pi}\,\Gamma(\alpha-\beta)\Gamma(\beta+1/2)}$$

and $[a]_+$ denotes

$$[a]_+ = \begin{cases} a, & \text{if} \quad a \geq 0 \\ 0, & \text{otherwise.} \end{cases}$$

• For $\alpha = \beta$, $\alpha > -\frac{1}{2}$, we have

$$W(x,y,z) = \begin{cases} \dfrac{\Gamma(\alpha+1)}{2\sqrt{\pi}\,\Gamma(\alpha+1/2)} \\ \qquad \times \dfrac{[1 - \text{ch}^2 x - \text{ch}^2 y - \text{ch}^2 z + 4\text{ch}^2 x\text{ch}^2 y\text{ch}^2 z]_+^{\alpha-1/2}}{(\text{sh}2x\text{sh}2y\text{sh}2z)^{2\alpha}}\ , \\ & \text{if } |x-y| < z < x+y \\ 0\ , & \text{otherwise} \end{cases}$$

(See [13], p. 94.)

Definition 2.2. (i) For all $x, y \in [0, +\infty[$, and f an even continuous function on $\mathbb{R}$, we put

$$\delta_x * \delta_y(f) = \int_{|x-y|}^{x+y} f(z) W(x,y,z) A(z) dz.$$

The mapping $*$ is called a generalized convolution product on $K = [0, +\infty[$.

(ii) We consider the identity mapping id as the involution on $K = [0, +\infty[$.

Theorem 2.2. *(i)* $(K, *)$ *is a hypergroup in the sense of Jewett called a Chébli–Trimèche hypergroup and denoted* $([0, +\infty[, *(A))$.

(ii) The Haar measure on $([0, +\infty[, *(A))$ *is given by* $A(x)dx$.

(iii) The dual space $\widehat{K}$ *can be parametrized by the disjoint union* $[0, +\infty[\cup i[0, \rho]$. *(See [1], [2], [17], [18], [20].)*

Examples.

Particular cases of Chébli–Trimèche hypergroups $([0, +\infty[, *(A))$ are Bessel–Kingman and Jacobi hypergroups which correspond respectively to the functions $A(x) = x^{2\alpha+1}$, $\alpha > -\frac{1}{2}$ and $A(x) = 2^{2(\alpha+\beta+1)}(\text{sh}x)^{2\alpha+1}(\text{ch}x)^{2\beta+1}$, $\alpha \geq \beta > -\frac{1}{2}$. (See [1], [18].)

Remark. The Chébli–Trimèche hypergroups $([0, +\infty[, *(A))$ are, for different functions A, equivalent to all regular hypergroups $([0, +\infty[, *)$ (see [10]).

Proposition 2.3. *There exists a function* $c : \mathbb{C} \to \mathbb{C}$ *satisfying*

$$\varphi_\lambda(x) = c(\lambda)\Phi_\lambda(x) + c(-\lambda)\Phi_{-\lambda}(x)$$

where Φ_λ *is the solution of the equation*

$$L_A u = -(\lambda^2 + \rho^2)u, \quad \lambda \in \mathbb{C}$$

such that

$$\Phi_\lambda(x) \sim e^{(i\lambda+\rho)x}, \quad x \to +\infty.$$

(i) The function $\lambda \to |c(\lambda)|^{-2}$ *is even on* $\mathbb{R}$, *nonnegative, continuous and*

$$c(-\lambda) = \overline{c(\lambda)}.$$

(ii) There exist positive constants N, k_1, k_2 *such that*

$$\forall\, \lambda \in \mathbb{R},\ |\lambda| \geq N,\ k_1|\lambda|^{2\alpha+1} \leq |c(\lambda)|^2 \leq k_2|\lambda|^{2\alpha+1}.$$

(iii) There exist $a, b \in \mathbb{R}$ *such that*

- *For* $\rho > 0$ *and* $\alpha > -\frac{1}{2}$, *we have*

$$|c(\lambda)|^2 \sim a\lambda^2, \quad \lambda \to 0.$$

- *For* $\rho = 0$ *and* $\alpha > 0$, *we have*

$$|c(\lambda)|^{-2} \sim b\lambda^{2\alpha+1}, \quad \lambda \to 0.$$

For (i) and (ii) see [2], [18]; for the proof of (iii) we have used Proposition 3.17 of [2], which is also Theorem 3.7.2 of [18].

Examples.

(i) In the case of the normalized Bessel function j_α we have

$$c(\lambda) = 2^\alpha \Gamma(\alpha+1) e^{-i(\alpha+1/2)\pi/2} \lambda^{-(\alpha+1/2)}.$$

(ii) In the case of the Jacobi function $\varphi_\lambda^{(\alpha,\beta)}$ we have

$$c(\lambda) = \frac{2^{\alpha+\beta+1-i\lambda}\Gamma(\alpha+1)\Gamma(i\lambda)}{\Gamma(\frac{1}{2}(\alpha+\beta+1+i\lambda))\Gamma(\frac{1}{2}(\alpha-\beta+1+i\lambda))}.$$

(See [14].)

Proposition 2.4. *(i) The Plancherel measure is given by*

$$d\pi(\lambda) = |c(\lambda)|^{-2} d\lambda.$$

(ii) We have

$$S = Supp\, \pi = [0, +\infty[.$$

(iii) The function $k(a)$ *defined by*

$$k(a) = \sup_{\lambda > 0} \frac{|c(\frac{\lambda}{a})|^{-2}}{|c(\lambda)|^{-2}}$$

is continuous on $[0, +\infty[$ *and satisfies the inequalities*

- *if* $\rho = 0$ *and* $\alpha > 0$: $\frac{M_1}{a^{2\alpha+1}} \leq k(a) \leq \frac{M_2}{a^{2\alpha+1}}$, *for all* $a > 0$

- *if* $\rho > 0$ *and* $\alpha > -\frac{1}{2}$: $\begin{cases} \dfrac{M_1}{a^{\gamma}} \le k(a) \le \dfrac{M_2}{a^{\gamma_0}}, & \text{if } a \ge 1 \\ \dfrac{M_1}{a^{\gamma_0}} \le k(a) \le \dfrac{M_2}{a^{\gamma}}, & \text{if } 0 < a < 1 \end{cases}$

where M_1 *and* M_2 *are positive constants,* $\gamma = \max(1, 2\alpha + 1)$ *and* $\gamma_0 = \min(1, 2\alpha + 1)$. *(For i) and ii) see [1], [18].)*

2.3. The hypergroup associated with the spherical mean operator

The spherical mean operator $\mathcal{R}$ is defined for a continuous function f on $\mathbb{R}^n$, even with respect to the first variable, by

$$\mathcal{R}(f)(r,y) = \int_{S^{n-1}} f(r\eta, y + r\xi)d\sigma_{n-1}(\eta,\xi); \ (r,y) \in \mathbb{R} \times \mathbb{R}^{n-1}$$

where S^{n-1} is the unit sphere $\{(\eta,\xi) \in \mathbb{R} \times \mathbb{R}^{n-1}/\eta^2 + \|\xi\|^2 = 1\}$ in $\mathbb{R}^n$ and σ_{n-1} is the surface measure on S^{n-1} of total mass equal to 1.

For $(\mu, \lambda) \in \mathbb{C} \times \mathbb{C}^{n-1}$, let $\varphi_{\mu,\lambda}$ be the function defined by

$$\forall \ (r,x) \in \mathbb{R} \times \mathbb{R}^{n-1}, \quad \varphi_{\mu,\lambda}(r,x) = \mathcal{R}(e^{-i\langle\lambda,\cdot\rangle} \cos \mu)(r,x)$$

where $\langle\cdot,\cdot\rangle$ is the scalar product on $\mathbb{C}^{n-1}$. Then we have

$$\varphi_{\mu,\lambda}(r,x) = e^{-i\langle\lambda,x\rangle} j_{\frac{n-2}{2}}(r\sqrt{\mu^2 + \lambda_1^2 + \cdots + \lambda_{n-1}^2})$$

with

$$j_{\frac{n-2}{2}}(rz) = \begin{cases} 2^{\frac{n-2}{2}} \Gamma\left(\frac{n}{2}\right) \dfrac{J_{\frac{n-2}{2}}(rz)}{(rz)^{\frac{n-2}{2}}}, & \text{if} \quad rz \ne 0 \\ 1, & \text{if} \quad rz = 0 \end{cases}$$

and $J_{\frac{n-2}{2}}$ is the Bessel function of first kind and order $\frac{n-2}{2}$.

The function $\varphi_{\mu,\lambda}$ satisfies the following properties.

Proposition 2.5. *(i) For every* $(\mu,\lambda) \in \mathbb{C} \times \mathbb{C}^{n-1}$, *the function* $(r,x) \to \varphi_{\mu,\lambda}(r,x)$ *is infinitely differentiable on* $\mathbb{R} \times \mathbb{R}^{n-1}$ *and even with respect to the first variable.*

(ii) For every $(r,x) \in \mathbb{R} \times \mathbb{R}^{n-1}$, *the function* $(\mu,\lambda) \to \varphi_{\mu,\lambda}(r,x)$ *is entire on* $\mathbb{C} \times \mathbb{C}^{n-1}$.

(iii) We have

$$|\varphi_{\mu,\lambda}(r,x)| \le 1, \quad \textit{for all} \quad (r,x) \in [0,+\infty[\times\mathbb{R}^{n-1}$$

if and only if (μ, λ) *belongs to the set*

$$\Gamma = [0, +\infty[\times \mathbb{R}^{n-1} \cup \{(i\mu, \lambda)/\mu \in \mathbb{R},\ \lambda \in \mathbb{R}^{n-1} \text{ and } |\mu| \leq \|\lambda\|\}.$$

(iv) For all $(r, x) \in \mathbb{R} \times \mathbb{R}^{n-1}$ *we have*

$$\varphi_{0,0}(r, x) = 1.$$

For every $(\mu, \lambda) \in \mathbb{C} \times \mathbb{C}^{n-1}$*, the function* $\varphi_{\mu,\lambda}$ *satisfies the product formula* $\forall\ (r, x), (s, y)$ *in* $[0, +\infty[\times \mathbb{R}^{n-1}$

$$\varphi_{\mu,\lambda}(r, x)\varphi_{\mu,\lambda}(s, y) = \frac{\Gamma\left(\frac{n}{2}\right)}{\sqrt{\pi}\,\Gamma\left(\frac{n-1}{2}\right)} \int_0^\pi \varphi_{\mu,\lambda}(\sqrt{r^2 + s^2 - 2rs\cos\theta}, x + y)(\sin\theta)^{n-2} d\theta\ .$$

(See [11], [15].)

Definition 2.3. (i) For all $(r, x), (s, y) \in [0, +\infty[\times \mathbb{R}^{n-1}$, and f a continuous function on $\mathbb{R}^n$, even with respect to the first variable, we put

$$\delta_{(r,x)} * \delta_{(s,y)}(f) = \frac{\Gamma\left(\frac{n}{2}\right)}{\sqrt{\pi}\,\Gamma\left(\frac{n-1}{2}\right)} \int_0^\pi f(\sqrt{r^2 + s^2 - 2rs\cos\theta},\ x + y)(\sin\theta)^{n-2} d\theta\ .$$

The mapping $*$ is called a generalized convolution product on $K = [0, +\infty[\times \mathbb{R}^{n-1}$.

(ii) The involution on K is defined by

$$(r, x)^- = (r, -x), \quad \text{for all} \quad (r, x) \in K.$$

Theorem 2.3. *(i)* $(K, *)$ *is a hypergroup in the sense of Jewett, called the hypergroup associated with the spherical mean operator* $\mathcal{R}$.

(ii) The Haar measure on K *is given by* $r^{n-1}drdx$.

(iii) The dual space $\widehat{K}$ *can be parametrized by the set* Γ.

Proposition 2.6. *(i) The Plancherel measure* $d\pi(\mu, \lambda)$ *is given by*

$$d\pi(\mu, \lambda) = \pi(\mu, \lambda) d\mu d\lambda$$

with

$$\pi(\mu,\lambda) = \frac{1}{2^{2n-3}\pi^{n-1}(\Gamma\left(\frac{n}{2}\right))^2}\,\{(\mu^2+\|\lambda\|^2)^{\frac{n-2}{2}}\mu\mathbf{1}_{[0,+\infty[\times\mathbb{R}^{n-1}}(\mu,\lambda)$$
$$+\,(\|\lambda\|^2-\mu^2)^{\frac{n-2}{2}}\mu\mathbf{1}_{[0,\|\lambda\|]\times\mathbb{R}^{n-1}}(\mu,\lambda)\}$$

where $\mathbf{1}_{[0,+\infty[\times\mathbb{R}^{n-1}}$ *and* $\mathbf{1}_{[0,\|\lambda\|]\times\mathbb{R}^{n-1}}$ *are the characteristic functions of the sets* $[0,+\infty[\times\mathbb{R}^{n-1}$ *and* $[0,\|\lambda\|]\times\mathbb{R}^{n-1}$.

(ii) We have

$$\begin{aligned} S &= \operatorname{Supp}\pi \\ &= [0,+\infty[\times\mathbb{R}^{n-1} \cup \Big\{(i\mu,\lambda)/\mu\in[0,+\infty[,\ \lambda\in\mathbb{R}^{n-1} \\ &\quad \textit{and } 0\le\mu\le\|\lambda\|\Big\}. \end{aligned}$$

(iii) The function $k(a)$ *defined on* $]0,+\infty[$ *by*

$$k(a) = \sup_{(\mu,\lambda)\in S\setminus\{0\}} \frac{\pi\left(\frac{\mu}{a},\frac{\lambda}{a}\right)}{\pi(\mu,\lambda)}$$

satisfies

$$k(a) = \frac{1}{a^{n-1}}, \quad \textit{for all} \quad a>0.$$

(For (i) and (ii), see [11], [15].)

3. Wavelets on hypergroups.

3.1. Wavelets on the hypergroup K

Definition 3.1. We say that a function g in $L^2(K,m)$ is a wavelet on K if there exists a constant C_g such that

(i) $0 < C_g < +\infty$

(ii) For almost all λ in S, we have

$$C_g = \int_0^\infty |\mathcal{F}(g)(a\lambda)|^2\frac{da}{a}.$$

Theorem 3.1. *Given* $a>0$ *and* g *a wavelet on* K *in* $L^2(K,m)$. *Then*

(i) The function $\lambda \to \mathcal{F}(g)(a\lambda)$ *belongs to* $L^2(S,\pi)$ *and we have*

$$\|\mathcal{F}(g)(a\cdot)\|_{2,\pi} \leq \left(\frac{k(a)}{a^n}\right)^{1/2} \|g\|_2$$

where

$$k(a) = \sup_{\lambda \in S\setminus\{0\}} \frac{\pi(\lambda/a)}{\pi(\lambda)}.$$

(ii) There exists a function g_a *in* $L^2(K,m)$ *such that*

$$\forall\, \lambda \in S, \quad \mathcal{F}(g_a)(\lambda) = \mathcal{F}(g)(a\lambda)$$

and we have

$$\|g_a\|_2 \leq \left(\frac{k(a)}{a^n}\right)^{1/2} \|g\|_2.$$

Proof. (i) From the conditions (1.5) and (1.6) we obtain by change of variables

$$\int_S |\mathcal{F}(g)(a\lambda)|^2 \pi(\lambda) d\lambda = \frac{1}{a^n} \int_S |\mathcal{F}(g)(\Lambda)|^2 \pi\left(\frac{\Lambda}{a}\right) d\Lambda$$

or

$$\int_S |\mathcal{F}(g)(\Lambda)|^2 \pi\left(\frac{\Lambda}{a}\right) d\Lambda \leq k(a) \int_S |\mathcal{F}(g)(\Lambda)|^2 \pi(\Lambda) d\Lambda.$$

We deduce the result from these relations and Theorem 1.2.

(ii) Theorem 1.2 gives the result. ∎

Proposition 3.1. *(i) There exists a unique function* α_t, $t > 0$, *in* $L^2(K,m)$ *such that*

$$\forall\, \lambda \in S, \quad \mathcal{F}(\alpha_t)(\lambda) = \exp[-t(\|\lambda\|^2 + \|\lambda_0\|^2)].$$

(ii) The function

$$g(x) = -\frac{d}{dt}\, \alpha_t(x) - \|\lambda_0\|^2 \alpha_t(x)$$

is a wavelet on K and we have

$$C_g = \frac{e^{-2t\|\lambda_0\|^2}}{8t^2}.$$

Proof. We deduce these results from condition (1.6), Theorem 1.2 and Definition 3.1. ■

Theorem 3.2. *Let g be a wavelet on K in $L^2(K, m)$. Then for all $a > 0$ and $b \in K$, the function*

$$g_{a,b} = T_b(g_a) \tag{3.1}$$

is a wavelet on K in $L^2(K, m)$ and we have

$$C_{g_{a,b}} = |\varphi_\lambda(b)|^2 C_g$$

where T_b, $b \in K$, are the generalized translation operators given in Definition 1.1.

Proof. From Theorem 3.1 the function g_a belongs to $L^2(K, m)$. Then from the properties of the operator T_b the function $g_{a,b}$ belongs to $L^2(K, m)$.

Using Corollary 1.1 (i) we obtain

$$\mathcal{F}(g_{a,b})(\lambda) = \varphi_\lambda(b)\mathcal{F}(g)(a\lambda), \quad \text{a.e. on} \quad S;$$

thus

$$C_{g_{a,b}} = |\varphi_\lambda(b)|^2 \int_0^\infty |\mathcal{F}(g)(a'\lambda)|^2 \frac{da'}{a'}$$

From Definition 3.1 we have

$$C_{g_{a,b}} = |\varphi_\lambda(b)|^2 C_g.$$

The properties of the function φ_λ and Theorem 1.2 imply that the function $g_{a,b}$ satisfies the conditions of Definition 3.1. ■

Proposition 3.2. *Let g be a wavelet on K in $L^2(K, m)$ such that the function $a \to g_a$ is continuous from $]0, +\infty[$ into $L^2(K, m)$. Then the function $(a, b) \to g_{a,b}$ is continuous from $]0, +\infty[\times K$ into $L^2(K, m)$.*

Proof. Let (a_0, b_0), (a, b) in $]0, +\infty[\times K$. From property (iv) of the translation operator T_b we deduce

$$\|g_{a,b} - g_{a_0,b_0}\|_2 \leq \|g_a - g_{a_0}\|_2 + \|T_b(g_{a_0}) - T_{b_0}(g_{a_0})\|_2.$$

We now obtain the result from the continuity of the function $a \to g_a$ from $]0, +\infty[$ into $L^2(K, m)$ and the fact that the function $b \to T_b(g_{a_0})$ is continuous from K into $L^2(K, m)$. ■

3.2. Continuous wavelet transform on K.

Definition 3.2. Let g be a wavelet on K in $L^2(K, m)$. We define the continuous wavelet transform on K for f in $L^2(K, m)$ by

$$\Phi_g(f)(a, b) = \int_K f(x)\bar{g}_{a,b}(x)dm(x), \quad \text{for all} \quad (a, b) \in]0, +\infty[\times K.$$

This relation can also be written in the form

$$\Phi_g(f)(a, b) = f * \bar{g}_a(b)$$

where $*$ is the convolution product given by Definition 1.2.

Proposition 3.3. *Let g be a wavelet on K in $L^2(K, m)$.*

(i) For f in $L^2(K, m)$ we have

$$\forall\ (a, b) \in]0, +\infty[\times K, \quad |\Phi_g(f)(a, b)| \leq \left(\frac{k(a)}{a^n}\right)^{1/2} \|f\|_2\|g\|_2.$$

(ii) For all f in $L^q(K, m)$, $q \in [1, 2]$, the mapping $b \to \Phi_g(f)(a, b)$ belongs to $L^r(K, m)$ with $r \in [1, +\infty]$ satisfying $\frac{1}{r} = \frac{1}{q} - \frac{1}{2}$, and we have

$$\|\Phi_g(f)(a, .)\|_r \leq \left(\frac{k(a)}{a^n}\right)^{1/2} |\ \|g\|_2\|f\|_q.$$

Proof. (i) From Definition 3.2 and Proposition 1.1 we have

$$\forall\ (a, b) \in]0, +\infty[\times K, \quad |\Phi_f(f)(a, b)| \leq \|f\|_2\|g_a\|_2,$$

and from Theorem 3.1 we have

$$\|g_a\|_2 \leq \left(\frac{k(a)}{a^n}\right)^{1/2} \|g\|_2;$$

thus

$$\forall\ (a,b)\in]0,+\infty[\times K,\quad |\Phi_g(f)(a,b)| \le \left(\frac{k(a)}{a^n}\right)^{1/2} \|f\|_2\|g\|_2.$$

(ii) We deduce the result from Definition 3.2 and Proposition 1.1. ∎

The following theorems are Plancherel and Parseval formulas for the continuous wavelet transform on K.

Theorem 3.3. *Let g be a wavelet on K in $L^2(K,m)$.*

(i) Plancherel formula for Φ_g.
For all f in $L^2(K,m)$ we have

$$\|f\|_2^2 = \frac{1}{C_g}\int_K \int_0^\infty |\Phi_g(f)(a,b)|^2 \frac{da}{a}\ dm(b).$$

(ii) Parseval formula for Φ_g.
For all f_1, f_2 in $L^2(K,m)$ we have

$$\int_K f_1(x)\overline{f_2(x)}\ dm(x) = \frac{1}{C_g}\int_K\int_0^\infty \Phi_g(f_1)(a,b)\overline{\Phi_g(f_2)(a,b)}\ \frac{da}{a}\ dm(b).$$

Proof. (i) From Definition 3.2 and the Fubini–Tonelli theorem we have

$$\frac{1}{C_g}\int_K\int_0^\infty |\Phi_g(f)(a,b)|^2\frac{da}{a}\ dm(b)$$
$$= \frac{1}{C_g}\int_0^\infty \left(\int_K |f * \bar{g}_a(b)|^2 dm(b)\frac{da}{a}\right).$$

We deduce from Corollary 1.2

$$\frac{1}{C_g}\int_K\int_0^\infty |\Phi_g(f)(a,b)|^2\frac{da}{a}\ dm(b)$$
$$= \frac{1}{C_g}\int_0^\infty \left(\int_S |\mathcal{F}(f)(\lambda)|^2|\mathcal{F}(g_a)(\lambda)|^2\pi(\lambda)d\lambda\right)\frac{da}{a}\ ;$$

then from Fubini–Tonelli's theorem we have

$$\frac{1}{C_g}\int_K\int_0^{\infty}|\Phi_g(f)(a,b)|^2\frac{da}{a}\,dm(b)$$
$$=\left(\int_S|\mathcal{F}(f)(\lambda)|^2\pi(\lambda)d\lambda\right)\left(\frac{1}{C_g}\int_0^{\infty}|\mathcal{F}(g_a)(\lambda)|^2\frac{da}{a}\right).$$

The result follows from Theorems 1.2, 3.1 and Definition 3.1.

(ii) We deduce the result from (i).

Corollary 3.1. *Let g be a wavelet on K in $L^2(K,m)$ such that for all $a>0$ the function g_a is positive. Then for all f in $L^2(K,m)$, we have the inversion formula for the transform Φ_g:*

$$f(\cdot)=\frac{1}{C_g}\int_K\int_0^{\infty}\Phi_g(f)(a,b)g_{a,b}(\cdot)\frac{da}{a}\,dm(b)$$

weakly in $L^2(K,m)$.

Proof. From Theorem 3.3 (ii) and Definition 3.2 we have for all h in $L^2(K,m)$

$$\int_K f(x)\overline{h(x)}\,dm(x)$$
$$=\frac{1}{C_g}\int_K\int_0^{\infty}\Phi_g(f)(a,b)\left(\int_K\overline{h(x)}\,g_{a,b}(x)dm(x)\right)\frac{da}{a}\,dm(b)\,,$$

but from Fubini–Tonelli's theorem and Theorem 3.3 (ii) we have

$$\int_K\int_0^{\infty}\int_K|\Phi_g(f)(a,b)||\bar{h}(b)|g_{a,b}(x)\frac{da}{a}\,dm(b)dm(x)$$
$$\leq\int_K\int_0^{\infty}\Phi_g(|f|)(a,b)\Phi_g(|\bar{h}|)(a,b)\frac{da}{a}\,dm(b)<+\infty.$$

Then from Fubini's theorem

$$\int_K f(x)\bar{h}(x)dm(x)$$
$$=\int_K\left(\frac{1}{C_g}\int_K\int_0^{\infty}\Phi_g(f)(a,b)g_{a,b}(x)\frac{da}{a}\,dm(b)\right)\bar{h}(x)dm(x)$$

and the result follows. ■

By Theorem 3.3 the continuous wavelet transform Φ_g on K is an isometry of the Hilbert space $L^2(K, m)$ into the Hilbert space $L^2(]0, +\infty[\times K, \frac{1}{C_g} \frac{da}{a} dm(b))$ (the space of square integrable functions on $]0, +\infty[\times K$ with respect to the measure $\frac{1}{C_g} \frac{da}{a} dm(b)$). For the characterization of the image of Φ_g we interpret the vectors $g_{a,b}$, $(a, b) \in]0, +\infty[\times K$, as a set of coherent states in the Hilbert space $L^2(K, m)$.

Definition 3.3. A set of coherent states in a Hilbert space $\mathcal{H}$ is a subset $\{g_\ell\}_{\ell\in\mathcal{L}}$ of $\mathcal{H}$ such that

(i) $\mathcal{L}$ is a locally compact topological space and the mapping $\ell \to g_\ell : \mathcal{L} \to \mathcal{H}$ is continuous.

(ii) There is a positive Borel measure $d\ell$ on $\mathcal{L}$ such that, for f in $\mathcal{H}$,

$$\|f\|^2 = \int_{\mathcal{L}} |(f, g_\ell)|^2 d\ell.$$

Theorem 3.4. *Let $\{g_\ell\}_{\ell\in\mathcal{L}}$ be a set of coherent states in a Hilbert space $\mathcal{H}$. Define the isometry Φ of $\mathcal{H}$ into $L^2(\mathcal{L}, d\ell)$ (the space of square integrable functions on $\mathcal{L}$ with respect to the measure $d\ell$) by*

$$\Phi(f)(\ell) = (f, g_\ell), \quad f \in \mathcal{H}.$$

Let F be in $L^2(\mathcal{L}, d\ell)$. Then F belongs to $\Phi(\mathcal{H})$ if and only if

$$F(\ell) = \int_{\mathcal{L}} F(\ell')(g_{\ell'}, g_\ell)d\ell'.$$

(See [9], p. 37–38.)

Now let $\mathcal{H} = L^2(K, m)$, $\mathcal{L} =]0, +\infty[\times K$. Choose a wavelet g on K in $L^2(K, m)$ such that the function $a \longrightarrow g_a$ is continuous from $]0, +\infty]$ into $L^2(K, m)$ and let $g_\ell = g_{a,b}$ be given by Theorem 3.2 if $\ell = (a, b) \in \mathcal{L}$. Then we have a set of coherent states. Indeed, (i) of Definition 3.3 is satisfied because of Proposition 3.2, and (ii) of Definition 3.3 is satisfied for the measure $\frac{1}{C_g} \frac{da}{a} dm(b)$ on $]0, +\infty[\times K$. (See Theorem 3.3 (i).)

Theorem 3.5. *Let Φ_g be the continuous wavelet transform on K, with g a wavelet on K in $L^2(K, m)$, such that the function $a \to g_a$ is* *continuous from $]0,+\infty[$ into $L^2(K,m)$. Let F be in $L^2(]0,+\infty[\times K,$ $\frac{1}{C_g}\ \frac{da}{a}\ dm(b))$. Then there exists a function f in $L^2(K,m)$ such that*

$$F = \Phi_g(f)$$

if and only if

$$F(a,b) = \frac{1}{C_g}\int_K\int_0^\infty F(a',b')\left(\int_K g_{a',b'}(x)\overline{g_{a,b}(x)}\ dm(x)\right)\frac{da'}{a'}\ dm(b').$$

Proof. Apply Theorem 3.4 with $\mathcal{H} = L^2(K,m)$, $\mathcal{L} =]0,+\infty[\times K$, coherent states $g_{a,b}$ and measure $d\ell$ given by $\frac{1}{C_g}\ \frac{da}{a}\ dm(b)$. ■

Theorem 3.6. *Let g be a wavelet on K in $L^2(K,m)$. For f a function in $C(K)\cap L^1(K,m)$ (resp. $C(K)\cap L^2(K,m)$) such that $\mathcal{F}(f)$ belongs to $L^1(S,\pi)$ (resp. $(L^1\cap L^\infty)(S,\pi)$) we have the following inversion formula for Φ_g:*

$$f(x) = \frac{1}{C_g}\int_0^\infty\left(\int_K \Phi_g(f)(a,b)g_{a,b}(x)dm(b)\right)\frac{da}{a} \tag{3.1}$$

where, for each $x\in K$, both the inner integral and the outer integral are absolutely convergent, but possibly not the double integral.

Proof. We put

$$i(a,x) = \int_K \Phi_g(f)(a,b)g_{a,b}(x)dm(b)$$

and

$$I(x) = \frac{1}{C_g}\int_0^\infty i(a,x)\frac{da}{a}; \tag{3.2}$$

we shall prove that for each $x\in K$, the integrals $i(a,x)$ and $I(x)$ are absolutely convergent and we have

$$I(x) = \int_S \mathcal{F}(f)(\lambda)\varphi_\lambda(x)\pi(\lambda)d\lambda.$$

(i) We suppose that $f\in C(K)\cap L^1(K,m)$ such that $\mathcal{F}(f)\in L^1(S,\pi)$. From Theorem 3.2 and Definition 3.2 we have

$$\Phi_g(f)(a,b)g_{a,b}(x) = f * \bar{g}_a(b)T_x(g_a)(b).$$

Proposition 1.1 and the properties of the operator T_x imply that the functions $b \to f * \bar{g}_a(b)$ and $b \to T_x(g_a)(b)$ belong to $L^2(K, m)$; then from Hölder's inequality the integral $i(a, x)$ is absolutely convergent. On the other hand, from Corollary 1.1 we have

$$\mathcal{F}(f * \bar{g}_a)(\lambda) = \mathcal{F}(f)(\lambda)\overline{\mathcal{F}(g_a)(\lambda)}, \quad \text{a.e. on} \quad S$$

and

$$\mathcal{F}(T_x(g_a))(\lambda) = \varphi_\lambda(x)\mathcal{F}(g_a)(\lambda), \quad \text{a.e. on} \quad S;$$

then using Theorem 1.2 we obtain

$$i(a, x) = \int_S \mathcal{F}(f)(\lambda)|\mathcal{F}(g_a)(\lambda)|^2 \varphi_\lambda(x)\pi(\lambda)d\lambda. \tag{3.3}$$

Thus

$$\frac{1}{C_g}\int_0^\infty |i(a, x)|\frac{da}{a} \leq \int_S |\mathcal{F}(f)(\lambda)| \left(\frac{1}{C_g}\int_0^\infty |\mathcal{F}(g_a)(\lambda)|^2 \frac{da}{a}\right) \pi(\lambda)d\lambda,$$

but from Definition 3.1 we have

$$\frac{1}{C_g}\int_0^\infty |\mathcal{F}(g_a)|^2 \frac{da}{a} = 1. \tag{3.4}$$

Then

$$\frac{1}{C_g}\int_0^\infty |i(a, x)|\frac{da}{a} \leq \|\mathcal{F}(f)\|_{1,\pi} < +\infty. \tag{3.5}$$

From this inequality we deduce that the integral $I(x)$ is absolutely convergent.

We now prove the relation (3.2). From the relation (3.3) we have

$$I(x) = \frac{1}{C_g}\int_0^\infty \left(\int_S \mathcal{F}(f)(\lambda)|\mathcal{F}(g_a)(\lambda)|^2 \varphi_\lambda(x)\pi(\lambda)d\lambda\right)\frac{da}{a}.$$

From the inequality (3.5) we deduce that we can apply Fubini's theorem. Then we have

$$I(x) = \int_S \mathcal{F}(f)(\lambda) \left(\frac{1}{C_g}\int_0^\infty |\mathcal{F}(g_a)(\lambda)|^2 \frac{da}{a}\right) \varphi_\lambda(x)\pi(\lambda)d\lambda;$$

we obtain the relation (3.2) from the relation (3.4). We deduce relation (3.1) from Theorem 1.2.

(ii) We suppose that $f \in C(K) \cap L^2(K,m)$ such that $\mathcal{F}(f) \in (L^1 \cap L^\infty)(S,\pi)$. From Theorem 1.4, the function $f * \bar{g}_a$ belongs to $L^2(K,m)$ and we have

$$\mathcal{F}(f * \bar{g}_a) = \mathcal{F}(f)\mathcal{F}(\bar{g}_a).$$

The same proof as for (i) gives the relation (3.2). We deduce relation (3.1) from Theorem 1.2. ■

Remark. Wavelets and continuous wavelet transforms on the double coset hypergroup $K = \mathcal{K}\backslash G/\mathcal{K}$, on Chébli–Trimèche hypergroups and on the hypergroup associated with the spherical mean operator are studied in [4], [15], [16], [17].

References

[1] **W.R. BLOOM–H. HEYER:** Harmonic analysis of probability measures on hypergroups, de Gruyter Studies in Mathematics, Vol. 20, Editors: H. Bauer–J.L. Kazdan–E. Zehnder, de Gruyter, Berlin–New York, 1994.

[2] **W.R. BLOOM–Z. XU:** The Hardy–Littlewood maximal function for Chébli–Trimèche hypergroups, Contemporary Math., Vol. 183, 1995, pp. 45–69.

[3] **C.K. CHUI:** An introduction to wavelets, Academic Press, 1992.

[4] **A. FITOUHI–K. TRIMECHE:** J. L. Lions transmutation operators and generalized continuous wavelet transform, Preprint. Faculty of Sciences of Tunis, 1995.

[5] **G.B. FOLLAND:** Real analysis: modern techniques and their applications, John Wiley & Sons, New York, Chichester–Brisbane–Toronto–Singapore, 1984.

[6] **R. GANGOLLI–V.S. VARADARAJAN:** Harmonic analysis of spherical functions and real reductive groups, Vol. 101, Springer–Verlag, Berlin–New York, 1988.

[7] **S. HELGASON:** Group and geometry analysis,, integral geometry, invariant differential operators and spherical functions, Academic Press, New York, 1984.

[8] **R.I. JEWETT:** Spaces with an abstract convolution of measures, Advances in Maths. 18, 1975, pp. 1–101.

[9] **T.H. KOORNWINDER:** Wavelets: An elementary treatment

of theory and applications, Edited by Tom H. Koornwinder, Series in Approximations and Decompositions, Vol. 1, World Scientific, 1993.

[10] **Y. MEYER:** Ondelettes et opérateurs I. Actualités Mathématiques, Hermann, Editeurs des Sciences et des Arts, 1990.

[11] **M.N. NESSIBI–L.T. RACHDI–K. TRIMECHE:** Ranges and inversion formulas for spherical means operator and its dual, J. Math. Anal. and Appl., Vol. 196, 1995, pp. 861–884.

[12] **A.L. SCHWARTZ:** Classification of one-dimensional hypergroups, Proceedings of the A.M.S., Vol. 103, No. 4, 1988, pp. 1073–1081.

[13] **K. TRIMECHE:** Transformation intégrale de Weyl et théorème de Paley–Wiener associés à un opérateur différentiel singulier sur $(0, +\infty)$, J. Math. Pures et Appl., Vol. 60, 1981, pp. 51–98.

[14] **K. TRIMECHE:** Transmutation operators and mean periodic functions associated with differential operators, Math. Reports, Vol. 4, No. 7, 1988, pp. 1–282; Harwood Academic Publishers, London–Paris–New York–Melbourne.

[15] **K. TRIMECHE:** Inversion of the spherical mean operator and its dual using spherical wavelets, Proceedings of the conference held in Oberwolfach, Oct. 23–29, 1994, Editor: H. Heyer, World Scientific, Singapore–New Jersey–London–Hong Kong, 1995.

[16] **K. TRIMECHE:** Continuous wavelet transforms on semisimple Lie groups and on Cartan motion groups, C. R. Acad. Sci. Canada, Vol. XVI, No. 4, 1994, pp. 161–165.

[17] **K. TRIMECHE:** Inversion of the J. L. Lions transmutation operators using generalized wavelets, Preprint. Faculty of Sciences of Tunis, 1995.

[18] **Z. XU:** Harmonic analysis on Chébli–Trimèche hypergroups, Ph.D. thesis, Murdoch University, Australia, 1994.

[19] **H.M. ZEUNER:** One dimensional hypergroups, Adv. Math., Vol. 76, No. 1, 1989, pp. 1–18.

[20] **H.M. ZEUNER:** Limit theorems for one-dimensional hypergroups, Habilitations–Schrift, Tübingen, 1990.

Faculty of Sciences of Tunis, Department of Mathematics, Campus 1060, Tunis, TUNISIA

Semigroups of Positive Definite Functions and Related Topics

Martin E. Walter

Dedicated to K.R. Parthasarathy

Abstract

Every locally compact group, G, has defined on it a semigroup of continuous, positive definite functions, $P(G)$. This semigroup, additionally equipped with a normalized, partially ordered, convex structure is a complete invariant of the underlying group. This semigroup has an identity and we investigate what it means to differentiate in the classical calculus sense at this identity. This leads us to the concept of a semiderivation. We are also naturally led to consider the cohomology of continuous, unitary representations of G, as well as the "screw functions" of J. von Neumann and I. J. Schoenberg, a Lévy-Khinchin formula, and a characterization of groups with property (T).

Introduction

This is a slightly modified version of a talk prepared for *The International Conference on Harmonic Analysis* held in Delhi, India, in December of 1995.

If G is a locally compact group, a complex-valued function, p, defined on G is said to be positive definite if for each choice of natural number $n = 1, 2, 3, \ldots$ and for each choice of n elements, $g_1, \ldots, g_n$,

---

1991 *Mathematics Subject Classification.* Primary 43A30, 22A30; Secondary 46L99.

Key words and phrases. C^*-algebra, convolution, completely bounded, duality, Fourier-Stieltjes algebra, locally compact group, positive definite function, matrix entry, unitary representation.

from the group, the n by n complex matrix

$$[p(g_i^{-1}g_j)] = \begin{bmatrix} p(g_1^{-1}g_1) & \cdots & p(g_1^{-1}g_n) \\ \vdots & \ddots & \vdots \\ p(g_n^{-1}g_1) & \cdots & p(g_n^{-1}g_n) \end{bmatrix}$$

is positive definite (note that the diagonal elements of this matrix are all $p(e)$, where e is the identity of the group G).

Let $P(G)$ be the set of continuous, positive definite functions on G. Let $P(G)_1 = \{p \in P(G) : p(e) \leqq 1\}$. Replacing $P(G)$ by $P(G)_1$ is what I call normalization. $P(G)_1$ is a convex set; if $p_1, p_2 \in P(G)_1$, then $\lambda p_1 + (1-\lambda)p_2 \in P(G)_1$ for $0 \leq \lambda \leq 1$. $P(G)_1$ is also partially ordered in the sense that if $p, q \in P(G)_1$, then $p \lesssim q$ if and only if $q - p$ is positive definite, i.e., $q - p \in P(G)$. In this paper we will use the symbol $\lesssim$ for the partial ordering relation on $P(G)$. Thus, for example, if $\mathbf{0}$ is the zero function, $\mathbf{0} \lesssim p$ for all $p \in P(G)$.

If $p, q \in P(G)_1$, then the pointwise product, pq, is also in $P(G)_1$. This follows from the fact that $pq(g) = p(g)q(g)$ for $g \in G$, and the fact that the Schur-Hadamard product of two positive definite matrices is again positive definite, see [7], Lemma D.12, p. 683. Note that if (a_{ij}) and (b_{ij}) are two matrices then the Schur-Hadamard product of these two matrices is $(a_{ij}) \circ (b_{ij}) = (a_{ij}b_{ij})$.

Thus, $P(G)_1$ is a partially ordered, convex semigroup. If $P(H)_1$ is another such partially ordered, convex semigroup for a group H, a map $\psi : P(G)_1 \to P(H)_1$ will be called an *isomorphism* in this context, if as a map ψ is one-to-one and onto, ψ is affine,[1] multiplicative, i.e., $\psi(pq) = \psi(p)\psi(q)$, for $p, q \in P(G)_1$, and order-preserving, i.e, $\lesssim$. The following result answers affirmatively a problem posed in [14] about 20 years ago.

Theorem 1. *$\psi : P(G)_1 \to P(H)_1$ is an isomorphism if and only if there exists a topological isomorphism $\theta : H \to G$.*

Proof. It is not hard to prove this theorem if you use [2]. For an interesting example with detailed structure worked out see [14]. Let us now give a reasonably detailed proof of Theorem 1. Given $\theta : H \to G$, it is quite straightforward to construct the isomorphism $\psi : P(G)_1 \to P(H)_1$. We leave this for the reader.

Assume ψ is an isomorphism of $P(G)_1$ onto $P(H)_1$. Without relabeling ψ it can be linearly extended to all of $P(G)$, by $\| p \| \, \psi(\frac{p}{\|p\|}) =$

[1]This means $\psi(\lambda p_1 + (1-\lambda)p_2) = \lambda\psi(p_1) + (1-\lambda)\psi(p_2)$, for $0 \leq \lambda \leq 1$ and $p_1, p_2 \in P(G)_1$.

$\psi(p)$, for all nonzero $p \in P(G)$.

The finite linear combinations of elements from $P(G)$ form a commutative Banach algebra, $B(G)$, called the Fourier-Stieltjes algebra of G, cf., [5], [13]. The Fourier-Stieltjes algebra of G is related to other algebras "on" G. Recall that there is associated with every locally compact group G the Banach $*$-algebra $L^1(G)$ of functions (modulo the usual almost everywhere equivalence relation), absolutely integrable with respect to left Haar measure. Addition in $L^1(G)$ is pointwise, convolution is the product and the involution is given by $f^{\#}(g) = \Delta^{-1}(g)f^{\flat}(g)$, where $f^{\flat}(g) = \overline{f(g^{-1})}$. The overbar, as usual, denotes complex conjugation.

This Banach $*$-algebra $L^1(G)$ has what is referred to as a universal, enveloping C^*-algebra, denoted $C^*(G)$, see [4], §13.9, for details. If things are getting a bit too abstract for the reader, take a break from this proof and see what all this means in the more concrete case of finite groups, discussed immediately after this proof.

The Fourier-Stieltjes algebra, $B(G)$, is isometrically linearly isomorphic with the linear (Banach) space of bounded linear functionals on $C^*(G)$. The double dual, [13], of $C^*(G)$, which is also the dual of $B(G)$, is a von Neumann algebra, $W^*(G)$. Thus $B(G)$ is the predual of $W^*(G)$.

With the above preliminaries out of the way, let us go back and linearly extend ψ from $P(G)$ to all of $B(G)$. First observe that every function $b \in B(G)$ can be written in a unique way as a linear combination of two "self-adjoint" elements, viz., $b = \frac{(b+b^{\flat})}{2} + i\frac{(b-b^{\flat})}{2i}$. Thus if we can linearly extend ψ to the self-adjoint elements of $B(G)$, we can extend ψ to all of $B(G)$. Now any self-adjoint element in $B(G)$ is of the form $p-q$, where $p, q \in P(G)$. Define ψ on this element as $\psi(p-q) = \psi(p) - \psi(q)$. We must verify that this extension is well-defined. But this follows, since if $p-q = p'-q'$, for $p, p', q, q' \in P(G)$, then $p+q' = p'+q \in P(G)$. Thus $\psi(p) + \psi(q') = \psi(p+q') = \psi(p'+q) = \psi(p') + \psi(q)$. Hence ψ is well-defined on self-adjoint elements, since $\psi(p) - \psi(q) = \psi(p') - \psi(q')$.

The above arguments can be applied to the inverse of ψ. Thus, ψ is a linear isomorphism of $B(G)$ onto $B(H)$. It is easy to verify that ψ is also a multiplicative map. Now, it is well-known that a positive linear mapping between preduals of von Neumann algebras is (norm) continuous, but a proof of this fact can be found in [2], Lemma 3.1. Applying this result to ψ and the inverse of ψ we have that both of these maps are norm continuous.

Taking the transpose of ψ, ${}^t\psi : W^*(H) \to W^*(G)$, we have a bicontinuous, positivity-preserving map that maps the spectrum of the commutative Banach algebra $B(H) \subset W^*(H)$ into the spectrum of the commutative Banach algebra $B(G) \subset W^*(G)$. Thus ${}^t\psi(e)$ is a positive element (with norm at most one) of the spectrum of $B(G)$, [13]. Thus by a result of Russo and Dye, cf. [11], Corollary 2.9, both ${}^t\psi$ and its inverse are norm decreasing maps. Thus ψ is an isometry. It thus follows by arguments in [13], pp. 29–32, that the groups G and H are topologically isomorphic. ■

Let us take a look at finite groups, which are trivially locally compact. There are some highly non-trivial and interesting aspects to the finite group case. For a finite group G, $L^1(G)$ and $C^*(G)$ coincide as algebras but have distinct norms.

Suppose $G = \{g_1, \ldots, g_n\}$ is a group with n elements, and suppose that $g_1 = e$, the identity. Write the group multiplication table symmetrically, and for simplicity suppose that the extreme upper left entry in this table is $g_1^{-1}g_1$. Thus

$$\begin{bmatrix} g_1^{-1}g_1 & \cdots & g_1^{-1}g_n \\ \vdots & \ddots & \vdots \\ g_n^{-1}g_1 & \cdots & g_n^{-1}g_n \end{bmatrix}$$

where the diagonal entries are all e.

Any function f on G can be represented as the n by n complex matrix obtained by applying the function f to each entry in the multiplication table for G; thus f is represented by the matrix $[f(g_i^{-1}g_j)]$. In this representation, convolution multiplication of functions becomes matrix multiplication and the involution becomes the usual conjugate transpose. Thus $L^1(G)$ and $C^*(G)$ can both be represented as the $*$-algebra of all complex n by n matrices arising from this construction. In the first case the norm is $\| f \|_1 = \sum_{i,j} |f(g_i^{-1}g_j)|$ and in the second case it is the operator norm of the matrix acting on $\mathbb{C}^n$.

One can find an n by n unitary matrix, U, called the Fourier–Plancherel unitary, such that $UC^*(G)U^* = \oplus_{i=1}^{d} M_{n_i}$, where M_{n_i} is the n_i by n_i complex matrices, i.e, $C^*(G)$ is the direct sum of matrix algebras.

If G is a general locally compact abelian group, then $C^*(G)$ is isomorphic with $C_o(\Gamma)$, the continuous, complex-valued functions that vanish at infinity on Γ, the dual group of continuous characters of G. In this case $P(G)_1$ identifies with the positive (regular, Borel) measures on Γ with total mass 1 or less, as well as the positive, linear functionals on $C^*(G)$ with norm one or less. This last identification holds in general, namely, $P(G)_1$ identifies with the positive linear functionals on $C^*(G)$

of norm one or less, via the equation

$$\langle p, \omega(f)\rangle = \int p(g)f(g)dg,$$

where $f \in L^1(G)$ and $\omega(f)$ is in $C^*(G)$, ω being the universal representation. Note that $P(G)_1$ also identifies with the positive linear functionals on $L_1(G)$ that arise via integration with respect to measurable functions. See [4], 13.4.5.

We have the following fundamental definitions (see [15]).

Definition 1. The (generalized) *translation* of $P(G)_1$ by $p_o \in P(G)_1$ is the map $p \in P(G)_1 \mapsto p_o p \in P(G)_1$, which is defined since $P(G)_1$ is a semigroup.

Definition 2. The *translate* of $a \in C^*(G)$ by $p \in P(G)_1$, denoted a_p, or $T_p a$, is the unique element in $C^*(G)$ that satisfies $\langle a_p, q\rangle = \langle a, pq\rangle$ for all $q \in P(G)_1$.

In [15] we show that T_p is a completely positive map of $C^*(G)$ into itself and hence satisfies the *Kadison–Cauchy–Schwarz* inequality:

$$T_p(x^*x) \geqq T_p(x)^*T_p(x) \text{ for } x \in C^*(G).$$

Note that $\geqq$ is the usual operator order in a C^* algebra. This inequality says that T_p defines a non-negative, bilinear form on $C^*(G)$ for each p, and it would be interesting (if we had time) to pursue some consequences of that fact.

We now introduce a simple-minded notion of differentiation at the identity in $P(G)_1$: If $\{p_n\}_{n\in\mathbb{N}} \subset P(G)_1$ is an arbitrary sequence converging weakly to $\mathbf{1}$, we consider the limit $\lim_{n\to\infty} \frac{p_n(g)-1}{\frac{1}{n}-0}$ for each $g \in G$. Thus

Definition 3. A *semitangent* vector at $\mathbf{1} \in P(G)_1$ is any continuous, complex-valued function ψ on G satisfying $\psi(g) = \lim_{n\to\infty} n(p_n(g)-1)$, for each $g \in G$ and some $\{p_n\} \subset P(G)_1$, $\{n\}$ the natural numbers. Note that given a semitangent ψ, defined by $\{p_n\}$ as above, it is implicit in the definition that $\psi(g)$ exists as a complex number for each $g \in G$, and hence that $\lim_{n\to\infty} p_n(g) = 1$ for all $g \in G$. If a semitangent at $\mathbf{1}$ is 0 at e, we say it is a normalized semitangent. We denote the set of all normalized semitangents by $N_0(G)$.

We are immediately led to the following characterization of semitangents on G.

Theorem 2. *Let ψ be a continuous, complex-valued function on G. Then ψ is a (normalized) semitangent at $\mathbf{1} \in P(G)_1$, i.e., $\psi \in N_0(G)$, if and only if $\psi(e) = 0, \psi(g^{-1}) = \overline{\psi(g)}$ for all $g \in G$, and for each choice of natural number n and each choice of n elements $g_1, \ldots, g_n$ from G the n by n matrix $\{\psi(g_i^{-1}g_j) - \psi(g_i^{-1}) - \psi(g_j)\}$ is positive hermitian, i.e.,*

$$(1) \qquad \sum_{i,j=1}^{n} \{\psi(g_i^{-1}g_j) - \psi(g_i^{-1}) - \psi(g_j)\}\lambda_i\overline{\lambda_j} \geq 0$$

for any choice of complex numbers $\lambda_1, \ldots, \lambda_n$.

Proof. The following calculation takes place in the weak closure of $C^*(G)$ in its universal representation. This closure is called the universal enveloping W^*-algebra, $W^*(G)$, and it contains a copy of G.

Suppose $\lim_{n\to\infty} n(p_n(g) - 1) = \psi(g)$ for all $g \in G$, where $p_n \in P(G)_1$ for all n and $p_n(e) = 1$ for all n. (If necessary, let $p'_n(g)\frac{p_n(g)}{p_n(e)}$. Then $\lim_{n\to\infty} n(p'_n - 1) = \psi - \psi(e)$ is a normalized semitangent.) Now consider the set of completely positive maps $\{T_{p_n}\}_{n=1}^{\infty}$ of $W^*(G)$ into itself. Let $x = \sum_{k=1}^{m} \lambda_k g_k$. Then

$$\begin{aligned}
&\lim_{n\to\infty} n\{T_{p_n}(x^*x) - (T_{p_n}x*)(T_{p_n}x)\} \\
&\quad = \lim_{n\to\infty} n\{T_{p_n}(x^*x) - x^*x + x^*x - (T_{p_n}x^*)x \\
&\qquad\quad + (T_{p_n}x^*)x - (T_{p_n}x^*)(T_{p_nx}) \\
&\quad = \lim_{n\to\infty} n(T_{p_n} - T_1)(x^*x) + \lim_{n\to\infty} n(T_1 - T_{p_n})x^*x \\
&\qquad\quad + \lim_{n\to\infty} T_{p_n}x^*n(T_1 - T_{p_n})x \\
&\quad = \lim_{n\to\infty} T_{n(p_n-1)}(x^*x) - \lim_{n\to\infty} (T_{n(p_n-1)}x)^*x \\
&\qquad\quad - \lim_{n\to\infty} (T_{p_n}xT_{n(p_n-1)}x).
\end{aligned}$$

Now for our choice of $x = \sum_{k=1}^{m} \lambda_k g_k$ we see that

$$\begin{aligned}
&\lim_{n\to\infty} T_{n(p_n-1)}\left(\sum_{k=1}^{m} \bar{\lambda}_k\lambda_i g_k^{-1}g_k\right) \\
&\qquad = \lim_{n\to\infty} \sum_{i,k=1}^{m} \bar{\lambda}_k\lambda_i n(p_n(g_k^{-1}g_i) - 1)g_k^{-1}g_i \\
&\qquad = \sum_{i,k=1}^{m} \bar{\lambda}_k\lambda_i\psi(g_k^{-1}g_i)g_k^{-1}g_i.
\end{aligned}$$

In a similar fashion,

$$\lim_{n\leftarrow\infty}\left(T_{n(p_n-1)}\left(\sum_{k=1}^{m}\lambda_k g_k\right)\right)^*\left(\sum_{i=1}^{m}\lambda_i g_i\right)=\sum_{i,k=1}^{m}\bar{\lambda}_k\lambda_i\psi(g_k^{-1})g_k^{-1}g_i\ ;$$

$$\lim_{n\to\infty}\left(T_{p_n}\sum_{k=1}^{m}\bar{\lambda}_k g_k^{-1}\right)\left(T_{n(p_n-1)}\left(\sum_{i=1}^{m}\lambda_i g_i\right)\right)=\sum_{i,k=1}^{m}\bar{\lambda}_k\lambda_i\psi(g_i)g_k^{-1}g_i\,.$$

For $x=\sum_{k=1}^{m}\lambda_k g_k$ the above limits can be viewed as being taken with respect to the norm in $W^*(G)$, e.g.,

$$\|T_{n(p_n-1)}g-\psi(g)g\|_{W^*(G)}=\|g\|_{W^*(G)}|n(p_n(g)-1)-\psi(g)|\to 0$$

as $n\to\infty$, for $g\in G$.

Now the norm limit of positive elements in $W^*(G)$ is positive; hence we have

$$\sum_{i,k=1}^{m}\bar{\lambda}_k\lambda_i\{\psi(g_k^{-1}g_i)-\psi(g_k^{-1})-\psi(g_i)\}g_k^{-1}g_i\geq 0.$$

Since $\mathbf{1}\in P(G)_1$ we have that

$$\left\langle\sum_{i,k=1}^{m}\bar{\lambda}_k\lambda_i\{\psi(g_k^{-1}g_i)-\psi(g_k^{-1})-\psi(g_i)\}g_k^{-1}g_i,\quad\mathbf{1}\right\rangle\geq 0,$$

thus

$$\sum_{i,k=1}^{m}\bar{\lambda}_k\lambda_i\{\psi(g_k^{-1}g_k)-\psi(g_k^{-1})-\psi(g_i)\}\geq 0$$

for each m, each choice of $g_1,\ldots,g_m$ in G and each choice of complex numbers $\lambda_1,\ldots,\lambda_m$.

We now turn to proving the converse. If the conditions on ψ of Theorem 2 really mean that ψ is a semitangent at $\mathbf{1}$ in $P(G)_1$, then in analogy with the theory of Lie groups it should not be unexpected that $\{e^{t\psi}\}_{t\geq 0}$ is a (pointwise continuous) one-parameter semigroup in $P(G)_1$. If we could indeed establish this, then ψ must be a semitangent, since $\psi=\lim_{t\to 0}(1/t)(e^{t\psi}-1)$, pointwise, and $e^{t\psi}\in P(G)_1$ for all $t\geq 0$.

We will thus show that a continuous function ψ satisfying equation (1) of Theorem 2 and $\psi(g^{-1})=\psi(g)$ for all $g\in G$, determines a one-parameter semigroup of continuous functions of positive type, $\{e^{t\psi}\}_{t\geq 0}$. Since $\psi(e) = 0$, this semigroup is in $P(G)_1$. The proof is straightforward. We merely must verify that for $t \geq 0$ that $e^{t\psi}$ is indeed of positive type, i.e.,

$$\sum_{i,j=1}^{n} \lambda_i \bar{\lambda}_i \exp(t\psi(g_j^{-1} g_i)) \geq 0$$

for each natural number n, each choice of complex numbers $\lambda_1, \ldots, \lambda_n$ and each choice of $g_1, \ldots, g_n$ in G.

Thus

$$\begin{aligned} &\sum_{i,j=1}^{n} \lambda_i \bar{\lambda}_j \exp(t\psi(g_j^{-1} g_i)) \\ &\quad = \sum_{i,j=1}^{n} \exp(t\{\psi(g_j^{-1} g_i) - \psi(g_j^{-1}) \\ &\qquad - \psi(g_i)\})[\lambda_i \exp(t\psi(g_i))][\bar{\lambda}_j \exp(\overline{t\psi(g_j)})] \\ &\quad \geq 0\,, \end{aligned}$$

since if $A = (\psi(g_j^{-1} g_i) - \psi(g_j^{-1}) - \psi(g_i))$ is a positive, hermitian, $n \times n$ matrix, the "Schur exponential" $e^{tA} \equiv I + tA + t^2 A \circ A + t^3 A \circ A \circ A + \cdots$ is also positive hermitian for $t \geq 0$. Note that we used the fact that $\psi(g^{-1}) = \overline{\psi(g)}$ for all $g \in G$. ■

Corollary. *Let ψ be a continuous function. Then ψ is a (normalized) semitangent vector at* **1** *in $P(G)_1$, i.e., $\psi \in N_0(G)$, if and only if $\{e^{t\psi}\}_{t\geq 0}$ is a one-parameter semigroup in $P(G)_1$, with $\{e^{t\psi(e)} = 1$ for all t.*

It is interesting to note that an alternative characterization of $\psi \in N_0(G)$ is as a positive form on an ideal in the $*$-algebra, $C_c(G)$, the continuous, complex-valued functions on G with compact support. Note that $C_c(G)$ is a $*$-subalgebra of $L^1(G)$. Thus $\psi \in N_0(G)$ if and only if $\psi(e) = 0$, $\psi^\flat = \psi$, and $\int_G \int_G \psi(y^{-1}x)\overline{f(y)} f(x) dy dx \geq 0$ for each $f \in C_c(G)$ such that $\int_G f(x) dx = 0$.

We are naturally led to the following definition.

Definition 4. A linear operator ∂ defined on a norm dense subspace of $C^*(G)$, which we denote $\mathrm{Dom}(\partial)$, with values in $C^*(G)$ is called a *semiderivation* on $C^*(G)$ if $x \in \mathrm{Dom}(\partial)$ implies $x^* \in \mathrm{Dom}(\partial)$ and

$$\partial(x^*x) \geq (\partial x^*)x + x^*\partial x$$

whenever $x \in \mathrm{Dom}(\partial)$ and $x^*x \in \mathrm{Dom}(\partial)$.

The prototypical example of a semiderivation is $D = \frac{d^2}{dx^2}$ on the real line. In this case we have

$$D(f\overline{f}) \geq (D\overline{f})f + \overline{f}(Df),$$

where $\geq$ stands for pointwise inequality of functions. In physics, D is an example of a completely dissipative operator. We remark that the equation defining semiderivations defines a nonnegative, bilinear form that warrants further investigation.

One might wonder if any higher derivative is a semitangent, or semiderivation, on say $\mathbb{R}^n$. The answer is no. The Fourier transform of D is $-x^2$. This is a special case of the Lévy-Khinchin formula on $\mathbb{R}^n$, which says that the semitangents on $P(\mathbb{R}^n)_1$ are given by

$$\begin{aligned} -\psi(y) &= c + iL(y) + Q(y) \\ &+ \int_{\mathbb{R}^n - \{0\}} \left[1 - exp(-i(x|y)) - \frac{i(x|y)}{1 + \| x \|^2} \right] \left[\frac{1 + \| x \|^2}{\| x \|^2} \right] d\mu(x) \end{aligned}$$

where $x, y \in \mathbb{R}^n$, $c \geq 0$, L is a continuous linear form, Q is a continuous, nonnegative quadratic form and μ is a nonnegative bounded measure on $\mathbb{R}^n - \{0\}$ such that the above integral converges, cf. [3].

Eric Larsen arrived at an analogous formula for the semitangents on $P(G)_1$ (see [8]). The main idea is to express a semitangent (or negative definite function) as a (weighted) integral over extreme rays of the convex cone of all negative definite functions, using the theory of Choquet.

Recall the following definition of a seminorm on a group.

Definition 5. A nonnegative, subadditive function ρ on G is called a *seminorm* on G, i.e., $\rho : G \to [0, \infty)$ and $\rho(gh) \leqq \rho(g) + \rho(h)$, where $g, h \in G$.

There are more seminorms than semitangents, as we can see from the following proposition. In fact, every non-compact, compactly generated group has unbounded seminorms, but some such groups (with property (T)) have only bounded semitangents, ψ. In the following G_d is the group G with the discrete topology.

Proposition 1. *If $\psi \in N_0(G_d)$, then $\rho_\psi = |\psi|^{\frac{1}{2}}$ is a seminorm on G.*

Proof. This result follows almost immediately by taking the determinant of the following matrix (which is positive definite):

$$\begin{bmatrix} \psi(e) - \overline{\psi(g)} - \psi(g) & \psi(h^{-1}g) - \overline{\psi(h)} - \psi(g) \\ \psi(g^{-1}h) - \overline{\psi(g)} - \psi(h) & \psi(e) - \overline{\psi(h)} - \psi(h) \end{bmatrix}.$$

■

J. von Neumann and I.J. Schoenberg studied the following problem, see [9], [12]. What (semi)metrics ρ on $\mathbb{R}$, the real line, exist such that $(\mathbb{R}, \rho)$ as a metric space is imbeddable isometrically in a (real) Hilbert space? In particular, they were interested in finding *screw functions* F on $\mathbb{R}$ so that $\rho(x, y) = F(x - y)$ for $x, y \in \mathbb{R}$. It turns out that this problem has an elegant solution for any locally compact group G. First let us make a formal definition.

Definition 6. A function F is a *screw function* on locally compact group G if (G, ρ) is, as a metric space, isometrically imbeddable into a (real) Hilbert space, where $\rho(g, h) = F(h^{-1}g)$ for $g, h \in G$.

Proposition 2. *A nonnegative, real-valued function F is a screw function on locally compact group G if and only if $-F^2 \in N_0(G)$.*

The proof of this result is not difficult, given the techniques now available, cf. [15]. What is interesting to note is that the screw functions of von Neumann and Schoenberg are so closely related to semitangents and, as it turns out, cohomology.

Cohomological considerations naturally lead us to the possible cocycles which can implement an isometric imbedding of G into some real Hilbert space. See [15] for details of how this is done.

Before stating our last theorem we need at least one more definition.

Definition 7. The function, **1**, which is identically 1 on G is a continuous, irreducible unitary representation of G of dimension 1 on Hilbert space. Let $\hat{G}$ be the collection of all unitary equivalence classes of continuous, topologically irreducible representations of G on Hilbert space with the Fell topology, see [4], §18.1.5. If $\{\mathbf{1}\}$ is an open set in $\hat{G}$, then G is said to have (Kazhdan's) *property (T)*.

One might guess that if G has property (T) that all derivatives at **1** would be trivial, and this is the case if trivial means bounded. The groups $SL(n, \mathbb{R})$ for $n \geq 3$ have property (T), for example, so such groups do exist.

In the following theorem $H^1(G, H(\pi))$ is the first (continuous) cohomology group of continuous, unitary representation π of G. If this does not mean anything to you, see [15]. For now let me close this presentation with a theorem from [1].

Theorem 3. *Let G be a locally compact, σ-compact group. The following are equivalent:*
(1) G has property(T);
(2) Every $\psi \in N_0(G)$ is bounded as a function on G;
(3) Every semiderivation ∂_ψ, induced on $C^*(G)$ by a $\psi \in N_0(G)$, is bounded as an operator on $C^*(G)$;
(4) $H^1(G, H(\pi)) = 0$ for all continuous unitary representations π of G.

Acknowledgements. We would like to thank the referee for paying careful attention to detail and making several helpful suggestions which improved the exposition of this paper.

Final Note. Since writing this paper, we have discovered an explicit process for recovering a (locally compact) group G from $P(G)_1$. This not only provides another proof of our Theorem 1 above, but it finally provides "the" nonabelian duality theory that most closely resembles the van Kampen-Pontriagin duality theory for (locally compact) abelian groups. This work is currently in preparation.

References

[1] C.A. Akemann and M.E. Walter, *Unbounded negative definite functions*, Canadian Journal of Mathematics **33** (1981), 862–871.

[2] Wolfgang Arendt and Jean DeCannière, *Order isomorphisms of Fourier algebras*, J. Funct. Anal. **50** (1983), 19–143.

[3] C. Berg and G. Forst, *Potential Theory on Locally Compact Abelian Groups*, Springer-Verlag, New York, 1975.

[4] J. Dixmier, *Les C*-algèbres et leurs représentations*, Cahiers Scientifiques, Fasc. 29, Gauthier-Villars, Paris, 1964.

[5] P. Eymard, *L'algèbre de Fourier d'un groupe localement compact*, Bull. Soc. Math. France **92** (1964), 181–236.

[6] Edwin Hewitt and Kenneth A. Ross, *Abstract Harmonic Analysis*, Vol. 1, Springer-Verlag, Berlin, 1963.

[7] Edwin Hewitt and Kenneth A. Ross, *Abstract Harmonic Analyis*, Vol. 2, Springer-Verlag, Berlin, New York, 1970.

[8] Eric Richard Larsen, *Negative Definite Functions on Locally Compact Groups*, Ph.D. Thesis, University of Colorado, Boulder, Colorado, 1982.

[9] J. von Neumann and I.J. Schoenberg, *Fourier Integrals and metric geometry*, Trans. Amer. Math. Soc. **50** (1941), 497–251.

[10] K. R. Parthasarathy, *Multipliers on locally compact groups*, Lecture Notes in Mathematics, No. **93**, Springer-Verlag, Berlin, New York, 1969.

[11] Vern I. Paulsen, *Completely bounded maps and dilations, No.* **146**, Pitman Research Notes in Mathematical Series, New York, 1986.

[12] I.J. Schoenberg, *Metric spaces and positive definite functions*, Trans. Amer. Math. Soc. **44** (1938), 522–536.

[13] Martin E. Walter, *W^*-algebras and nonabelian harmonic analysis*, J. Functional Analysis **11** (1972), 17–38.

[14] Martin E. Walter, *Duality Theory for Nonabelian Locally Compact Groups*, Symposia Mathematica, **XXII**, (1977), 47–59.

[15] Martin E. Walter, *Differentiation on the Dual of a Group: An Introduction*, Rocky Mountain Journal of Mathematics **12** (1982), 497–536.

Email: walter@boulder.colorado.edu; Department of Mathematics, Campus Box 395, University of Colorado, Boulder, Colorado, U.S.A. 80309

Characters, Bi-Modules and Representations in Lie Group Harmonic Analysis

N.J. Wildberger

Abstract

This paper is a personal look at some issues in the representation theory of Lie groups having to do with the role of commutative hypergroups, bi-modules, and the construction of representations. We begin by considering Frobenius' original approach to the character theory of a finite group and extending it to the Lie group setting, and then introduce bi-modules as objects intermediate between characters and representations in the theory. A simplified way of understanding the formalism of geometric quantization, at least for compact Lie groups, is presented, which leads to a canonical bi-module of functions on an integral coadjoint orbit. Some meta-mathematical issues relating to the construction of representations are considered.

1. Introduction

This paper describes an approach to non-commutative harmonic analysis on a Lie group G which is based on an old idea of Frobenius. We discuss the possible role of commutative hypergroups (in the same vein as [19], [24]), introduce G bi-modules and consider a computational approach to the construction of irreducible representations. Most of the ideas are of an elementary nature. Along the way certain amusing but perhaps unsettling philosophical points are raised.

This is a personal approach– the opinions expressed are those of the author, and occasionally diverge from the mainstream view. While there is then the increased likelihood of saying something foolish, there is the advantage of encouraging discussion. Good natured debate is a sign of vigour in a theory, and it is in this spirit that I offer this paper.

Harmonic analysis on a non-commutative finite group G was initiated one hundred years ago by G. Frobenius in a paper entitled 'Über Gruppencharaktere' [8]. In this fundamental work Frobenius introduced the idea of a character and computed some character tables.

Nowadays we consider characters useful objects associated to representations, which were not mentioned in Frobenius' 1896 paper. They were introduced by him shortly thereafter when the modern definition appears: one starts with a representation $\pi : G \to Gl(n)$ and defines the associated character to be the function $\chi(g) = \text{tr } \pi(g)$. Our attitude is captured by the following quote of G. Mackey [13].

> Frobenius' original definition of character was a complicated one, which emerged from his analysis of Dedekind's problem. A year later, however, he showed that his definition is equivalent to another that is much simpler and more natural.

In other words, the fundamental question today is not

Question 1. What are all the characters of a group G?

as it was for Frobenius in 1896 but rather

Question 3. What are all the equivalence classes of irreducible representations of G?

In my opinion, Frobenius' original point of view towards characters deserves to be reconsidered as possibly more fundamental than the subsequent modern approach. From this standpoint, Question 1 becomes the natural starting point for harmonic analysis on a group G and Question 3 logically follows it. Between these is another reasonable question which seems to have attracted little attention.

Question 2. What are all the equivalence classes of irreducible G bimodules?

Frobenius' 1896 approach can be extended to other families of groups by utilizing the concept of a locally-compact commutative hypergroup and possible generalizations. The central object of interest is the *class hypergroup* $\mathcal{K}(G)$ of conjugacy classes of a group G.

For a compact Lie group G, $\mathcal{K}(G)$ is a locally compact, compact hypergroup in the sense of Dunkl [7], Jewett [11] and Spector [15]. There is a basic relation between this hypergroup and the hypergroup of adjoint orbits $\mathcal{K}(\mathfrak{g})$ in the Lie algebra $\mathfrak{g}$ of G given by the *wrapping map* (see [6]). From this vantage point many important aspects of harmonic analysis are seen in a new and simplified light. In particular, the classification of irreducible representations by highest weights or integral coadjoint orbits, the Kirillov character formula [12], a formula of Harish-Chandra on

G-invariant differential operators, the Duflo isomorphism, the Poisson-Plancherel formula of M. Vergne [18] and the identity of Thompson [17], amongst other things, can be explained both naturally and easily.

For non-compact Lie groups, the definition of $\mathcal{K}(G)$ is more subtle and problematic. At this point, little is known about the right way of proceeding in general although there is evidence that a suitable form of the wrapping map should apply in some situations. What seems clear, however, is that the present theory of locally compact hypergroups is too narrow to accomodate objects which arise as quotients of non-compact group actions. We need to understand hypergroups based on spaces which are not necessarily Hausdorff or locally compact. Its seems a project of some importance to establish such a theory and apply it to the harmonic analysis of non-compact Lie groups. Such a development is also likely to be of interest to mathematical physicists.

In Section 2, we define hypergroups, describe Frobenius' original approach to characters, and indicate how to extend this to a compact Lie group. In Section 3, we introduce the notion of G bi-modules and their potential role as intermediate objects between characters and representations. In Section 4 the natural occurence of such G bi modules for a compact Lie group G as spaces of functions on coadjoint orbits and the connection with moment maps and geometric quantization is described. The last section raises the possibility that the problem of 'constructing' representations of groups has been obscured by the lack of precision in usage of words like 'construct'. I propose a reasonably concrete meaning for this term and suggest that the task of constructing the irreducible representations of a compact Lie group is by no means completed.

Some problems of interest are scattered through the paper.

2. Hypergroups and Frobenius' approach to characters

The concept which clarifies Frobenius' paper is that of a *finite commmutative hypergroup.* This notion is almost implicit in Frobenius' 1896 work. In retrospect, it seems curious that in a century of mathematics oriented towards abstract algebra, this important theory has been developed only recently. Since we will be interested in applications to Lie groups we give a more general definition without a precise discussion of the topologies involved- see [11] or [4] for more detail.

Definition 1. *A* locally compact commutative hypergroup *is a locally compact space $\mathcal{K}$ for which the Borel measures $M(\mathcal{K})$ form a $*$-algebra* *satisfying essentially the following axioms.*

(1) (Closure) The product of Dirac delta functions $\delta_x \times \delta_y$*, for* $x, y \in \mathcal{K}$*, is always a compactly supported probability measure which varies continously with* x *and* y*.*

(2) (Associativity) The algebra $M(\mathcal{K})$ *is associative.*

(3) (Existence of an Identity) There exists an element $e \in \mathcal{K}$ *such that* δ_e *is the identity.*

(4) (Existence of Inverses) For every $x \in \mathcal{K}$ *there exists a unique element* $x^* \in \mathcal{K}$ *such that* e *is contained in the support of the measure* $\delta_x \times \delta_{x^*}$*. Furthermore* $(\delta_x)^* = \delta_{x^*}$*.*

(5) (Commutativity) The algebra $M(\mathcal{K})$ *is commutative.*

In the case of a finite set $\mathcal{K} = \{c_0, c_1, \cdots, c_n\}$, this definition coincides, once we identify measures and functions, with the definition given in the paper [16] in this same volume. A finite commutative hypergroup is as close to being a commutative group as possible given that we only require the product of two elements to be a probability distribution of elements; in particular a commutative group is also a commutative hypergroup. The notions of character, duality and Fourier transform for commutative groups extend to finite commutatve hypergroups once we consider also somewhat more general objects called *signed hypergroups* which involve negative probabilities.

Frobenius' original approach may now be restated into modern language as follows. For any non-commutative finite group G, there is an associated finite commutative hypergroup $\mathcal{K}(G)$, called the *class hypergroup* of G, obtained from convolving G-invariant probability measures supported on conjugacy classes. The characters of the hypergroup $\mathcal{K}(G)$ form a signed hypergroup $\mathcal{K}(G)^\wedge$ which happens to be a hypergroup. The elements of this dual hypergroup are functions on $\mathcal{K}(G)$ and so can be naturally interpreted as central functions on G. They are precisely the irreducible characters in the usual sense, except that they have been normalized to have value 1 at the identity.

A number of facts about characters of finite groups such as the orthogonality relations and integrality of character values are special cases of more general facts which hold for finite commutative hypergroups (for some deeper examples, see [2], where the terminology 'table algebras' are used). This is useful since there are many examples of finite commutative hypergroups which arise outside of group theory, for example in the theory of distance regular graphs, association schemes, conformal field theory, cyclotomy, inclusions of Von Neumann algebras etc., see [20].

Summarizing, we may say that the problem of determining the characters of a non-commutative group G is essentially a problem of commutative harmonic analysis; analysis on the associated class hypergroup $\mathcal{K}(G)$.

Let us now try to generalize this approach to the case of G a compact Lie group. Let $\mathcal{K} = \mathcal{K}(G)$ be the set of conjugacy classes of G, which we may view as the quotient of G with respect to the conjugation action on itself. Measures on $\mathcal{K}(G)$ form an algebra under convolution induced by the convolution algebra of central measures on G. In this correspondence, delta functions on $\mathcal{K}(G)$ are associated to invariant probability measures on conjugacy classes.

A character of $\mathcal{K}(G)$ is a bounded continuous function $\chi : \mathcal{K}(G) \to \mathbb{C}$ such that

(1) $\chi(x)\chi(y) = \int_{\mathcal{K}(G)} \chi(z) d(\delta_x \times \delta_y)(z)$ for all $x, y \in \mathcal{K}$

(2) $\chi(x^*) = \overline{\chi(x)}$ for all $x \in \mathcal{K}$.

The set of characters of a hypergroup $\mathcal{K}$ is denoted by $\mathcal{K}^\wedge$. A character of $\mathcal{K}(G)$ lifts via the quotient map $p : G \to \mathcal{K}(G)$ to a function on G invariant on conjugacy classes.

Definition 2. *Any function on G obtained this way from a character of* $\mathcal{K}(G)$ *is called an* irreducible normalized character *of G.*

It is a consequence of a well known theorem of Weyl that this notion is exactly the same as the usual one; that is, an irreducible normalized character is exactly a function of the form $\chi(g) = \frac{1}{n}$ tr $\pi(g)$ for some irreducible representation $\pi : G \to Gl(n)$.

Our strategy towards understanding harmonic analysis on a compact Lie group G is now the following. The first step is to understand $\mathcal{K}(G)$ and its hypergroup structure– this is the basic object. The next step is to determine the characters of $\mathcal{K}(G)$ and the hypergroup structure of $\mathcal{K}(G)^\wedge$. The next step is to construct as explicitly as possible the irreducible bi-modules of G and the irreducible representations of G.

The first two steps can be accomplished by turning our attention to the Lie algebra $\mathfrak{g}$ and to certain hypergroup structures on it and its dual $\mathfrak{g}^*$. These hypergroup structures are special cases of a general phenomenon: for any linear action of G on a vector space V, the space of orbits of G on V carries a hypergroup structure obtained by convolving G-invariant probability measures in $\mathfrak{g}$.

The resulting orbit hypergroup $\mathcal{K}(V; G)$ has dual the orbit hypergroup $\mathcal{K}(V^*; G)$. Here G acts on the dual vector space as follows:

$$g \cdot f(v) = f(g^{-1} \cdot v) \text{ for all } g \in G, f \in V^*, v \in V.$$

Furthermore the pairing between the G orbit $\mathcal{U}$ of V and the G orbit $\mathcal{O}$ of V^* is given by:

$$\langle \mathcal{U}, \mathcal{O} \rangle = \int_{\mathcal{U}} \int_{\mathcal{O}} e^{if(v)} d\mu_{\mathcal{O}}(f) d\mu_{\mathcal{U}}(v) \tag{1}$$

where $d\mu_{\mathcal{O}}$ and $d\mu_{\mathcal{U}}$ are the G-invariant probability measures on $\mathcal{O}$ and $\mathcal{U}$.

Since G acts on $\mathfrak{g}$ and on $\mathfrak{g}^*$ by the adjoint and coadjoint actions respectively, we have hypergroups $\mathcal{K}(\mathfrak{g}\,; G)$ and $\mathcal{K}(\mathfrak{g}^*; G)$ which are in duality.

To understand the connection between the class hypergroup $\mathcal{K}(G)$ and the *adjoint hypergroup* $\mathcal{K}(\mathfrak{g}\,; G)$ we now introduce the wrapping map Φ from distributions of compact support on $\mathfrak{g}$ to distributions on G. The consideration of this map is motivated by the work of Harish-Chandra, Helgason, Kashiwara- Vergne, and Duflo. Let j denote a suitably chosen square root of the Jacobian of the exponential map $\exp : \mathfrak{g} \to G$. This function is analytic on $\mathfrak{g}$, has value 1 at 0 and is G-invariant (with respect to the adjoint action). For a function φ on G let $\tilde{\varphi} = \varphi \circ \exp$ be its lift to $\mathfrak{g}$. For a distribution μ of compact support on $\mathfrak{g}$, define the distribution $\Phi(\mu)$ on G by

$$\langle \Phi(\mu), \varphi \rangle = \langle \mu, j\tilde{\varphi} \rangle$$

for any $\varphi \in C^\infty(G)$.

The following result, which we call the *wrapping theorem*, (see [6]) is a generalization of results of Harish-Chandra, Duflo and I. Frenkel.

Theorem 3. *Let μ and ν be two G-invariant distributions of compact support on $\mathfrak{g}$. Then*

$$\Phi(\mu) * \Phi(\nu) = \Phi(\mu * \nu)$$

where the convolution on the left is group convolution on G and the convolution on the right is Euclidean convolution in $\mathfrak{g}$.

The wrapping theorem allows us to relate the structures of the hypergroups $\mathcal{K}(G)$ and $\mathcal{K}(\mathfrak{g}\,; G)$. This explains why Kirillov's orbit theory of representations works. Coadjoint orbits and representations are intimately related since the former determine characters of the adjoint hypergroup and the latter determine characters of the class hypergroup. The character formula of Kirillov follows from the pairing of adjoint and coadjoint orbits. For a more extensive discussion of the applications of this point of view to compact Lie groups, see [6] , [24].

How does any of this generalize to non-compact Lie groups? This is not at all so clear, since the notion of a G-invariant probability measure on a conjugacy class in general is not defined, and even if it were, how would one convolve two such things? There are some indications that an appropriate theory of means could be used for some groups at least (see [19]). The fact that much of Kirillov theory extends to non-compact groups suggests that such an approach ought to exist. We know what the answer is– the theory of characters of non-compact Lie groups as developed in the nilpotent case by Kirillov and in the semisimple case by Harish-Chandra and others; the question is – how to recover these results from a hypergroup approach? In other words, how can one obtain the results of representation theory on characters for non-compact Lie groups without mentioning representations? The following is thus a key project.

Problem 1. Develop a more general theory of hypergroups into which the class hypergroups of non-compact Lie groups may fit.

3. G bi-modules

In this section we introduce a family of objects which perhaps can play a role in harmonic analysis somewhere between characters and representations. The notion of a bi-module is familiar enough in other areas of mathematics, for example the theory of Von Neumann algebras where it has been introduced by Connes.

Let G be a Lie group; that is a smooth manifold with a compatible smooth group structure.

Definition 4. *A* left-action *of G on a finite dimensional vector space V is an assignment to each $g \in G$ and $v \in V$ an element $g \cdot v \in V$ such that*

(1) $g \cdot v$ varies smoothly with g and v
(2) $g \cdot v$ varies linearly with v
(3) $g \cdot (h \cdot v) = (gh) \cdot v$ for all $g, h \in G, v \in V$.

We will also say that G *left-acts* on V, or that V is a G *left-module.*

Definition 5. *A* right-action *of G on a finite-dimensional vector space V is an assignment to each $g \in G$ and to each $v \in V$ an element $v \cdot g \in V$ such that*

(1) $v \cdot g$ varies smoothly with g and v
(2) $v \cdot g$ varies linearly with v
(3) $(v \cdot g) \cdot h = v \cdot gh$ for all $g, h \in G, v \in V$.

In this case we say that V is a G *right-module.* Notice that if V is say a G left-module then V^* can be made into a G right-module by defining

$$(f \cdot g)\ (v) = f(g \cdot v) \text{ for all } f \in V^*, g \in G, v \in V.$$

Definition 6. *A vector space W is a G* bi-module *if it is both a G left-module and a G right-module and if these actions are* compatible, *that is if*

$$g \cdot (w \cdot h) = (g \cdot w) \cdot h \text{ for all } g, h \in G, w \in W.$$

Definition 7. *A G bi-module W is* irreducible *if there is no proper subspace $W_0 \subset W$ which is stable under both actions.*

Definition 8. *A G bi-module W is* symmetric *if there exists an involution (a linear map whose square is the identity) $^* : W \to W$ such that*

$$(g \cdot (w \cdot h))^* = (h^{-1} \cdot w^*) \cdot g^{-1}$$

If V is a G left-module then V^* is a G right-module and the tensor product $V \otimes V^*$ is a G bi-module in the obvious way. If in addition we are given a map $^* : V \to V^*$ such that

$$(g \cdot v)^* = v^* \cdot g^{-1}$$

then $V \otimes V^*$ becomes a symmetric G bi-module by defining

$$(v \otimes u^*)^* = u \otimes v^*.$$

This is the situation when a G left-action on V preserves a bilinear symmetric form (,) on V.

Problem 2. For a given group G, construct its irreducible G bi-modules (up to the natural notion of equivalence), its irreducible symmetric bi-modules etc.

Remark 1. The requirement of smoothness in the definition of a left-action is not standard. In the literature smoothness is usually replaced by continuity. The reasons are perhaps that

(a) for compact Lie groups the two notions happen to coincide, and

(b) requiring only continuity allows an immediate extension of the definition to infinite dimensional vector spaces.

Quite frankly, this seems to me an unfortunate sleight of hand. Functorial and aesthetic considerations urge us to respect the smooth structure of a Lie group. This means taking care to investigate the meaning of smooth structures on infinite dimensional vector spaces before we extend our definitions to that realm. Since representation theory is of such interest and usefulness, it is appropriate that the basic definitions be natural and functorial.

Remark 2. If one takes a Lie group to mean analytic manifold with analytic group operations, then the above definitions should be modified appropriately. Perhaps harmonic analysis on a smooth Lie group is quite a different subject from harmonic analysis on an analytic Lie group. Note that on the latter, distribution theory is sometimes not naturally available– on a non-compact analytic Lie group for example, there are no naturally occurring functions of compact support!

Remark 3. The terminology of left-action and right-action motivates us to constantly acknowledge the presence of any assymetry which we introduce into the situation. This is a another 'philosophical' point which I believe is of some general usefulness. It is most pleasant when our basic concepts are free of arbitrary bias, even the seemingly innocent one of left vs right. In this context, does anyone know how to define the Lie algebra of a Lie group in a completely symmetric way?

Nevertheless, we will continue to follow standard usage and interchange the terms left-action and representation.

Why should we consider G bi-modules? The reason is that they occur naturally. The most obvious representation of any finite group is the regular representation on the space of all functions on the group. This is of course naturally a G bi-module, since we may translate functions on the left or on the right. As a G bi-module, this space decomposes into irreducible G bi-modules, one for each of the irreducible representations of the group. Each such constituent W is a two sided ideal of the convolution algebra of all functions and contains the character of the representation as an minimal idempotent. As a left-module, W decomposes into dim V copies of an irreducible left-module V, but the decomposition is not canonical. This is a recurring theme in harmonic analysis and quantization theory; the canonical 'square' object which has no natural decomposition into 'left' or 'right' objects. Here we are examining the 'square' objects in their own right.

Given a representation of G on a space V, there is an obvious G

bi-module structure on $W = \text{End } V \simeq V \otimes V^*$, which is in addition an algebra. Equivalently we may obtain W by considering the space of all matrix coefficients of the representation. The other direction, from W to V, is not at all so easy and is rarely canonical– which is why the bi-module W perhaps qualifies as a simpler object than the representation V.

4. Geometric quantization revisited

As an illustration of the occurrence of G bi-modules in harmonic analysis, let us return to the situation of a compact Lie group G and consider the method of geometric quantization which associates to an integral coadjoint orbit $\mathcal{O}$ an irreducible representation π of G, that is, the Borel–Weil theorem. We briefly review this theory (see [10], [14], [25],) and then sketch a simpler way of understanding it by considering moment maps of representations. The possibility of such a simplification was suggested to the author by a conversation with I. Frenkel. The presence of a canonical G bi-module of functions on $\mathcal{O}$ follows from this approach.

The assumption that $\mathcal{O}$ is integral means that it carries a complex structure (not unique) such that there exists a holomorphic Hermitian line bundle L over $\mathcal{O}$ with connection whose curvature is equal to the canonical symplectic form on $\mathcal{O}$ (all coadjoint orbits are symplectic manifolds). There is a formula involving the covariant derivative of the connection that shows that the Lie algebra $\mathfrak{g}$ acts naturally on the space of all sections of this bundle. Equivalently one may rephrase this in terms of a circle bundle over $\mathcal{O}$ and spaces of functions on this bundle that transform correctly under the S^1 action (see [14] for more details). The irreducible module associated to the orbit is obtained by taking the submodule of holomorphic sections. All irreducible representations of G occur in this way.

The procedure is admittedly somewhat magical, but there is a more conceptual way of understanding it in which the objects and constructions are more natural. The key is to emphasize the inverse procedure of *dequantization* (see [5], [21]).

Dequantization in this context means going backwards from an irreducible unitary representation $\pi : G \to U(V)$ to the coadjoint orbit $\mathcal{O}$ and its associated geometric objects. Here $U(V)$ is the group of unitary operators on some complex inner product space V. Let Ω be the unit sphere of V. We consider the map $\phi : \Omega \to \mathfrak{g}^*$ defined by

$$\phi(v)(X) = \frac{1}{i} \langle d\pi(X)v, v\rangle$$

for $v \in \Omega$ and $X \in \mathfrak{g}$. The map ϕ is the composition of the projection of Ω onto the projective space PV of V and the moment map of the representation ([9], [21], [22], [23]).

The image of ϕ is a G-invariant compact subset of $\mathfrak{g}^*$ whose convex hull has extremal set a single coadjoint orbit $\mathcal{O}$. In fact most of the time the image of ϕ is convex , see [1] and [23] for the exact statement. In any case the preimage of this extremal orbit $\mathcal{O}$ is a single G orbit $\mathcal{M}$ in Ω and $\phi : \mathcal{M} \to \mathcal{O}$ is a circle bundle; the S^1 action is the restriction to $\mathcal{M}$ of the natural multiplicative action of S^1 on V. The orbit $\mathcal{O}$ is the same coadjoint orbit that geometric quantization utilizes in the construction of π; $\mathcal{M}$ is actually the orbit of the highest weight vector of the representation and the circle bundle over $\mathcal{O}$ is the same as that constructed by geometric quantization. The holomorphic and Hermitian structure of the associated line bundle L follow directly from the holomorphic and Hermitian structure of the space V.

There is furthermore a natural way of assigning to any vector $v \in V$ a function f_v on $\mathcal{M}$ by the rule $f_v(w) = \langle v, w\rangle$ for $w \in U$. This function does not push down to $\mathcal{O}$ but nevertheless has exactly the correct transform properties to identify it with a section of the corresponding line bundle L. This realizes V as a space of sections of L on which G acts.

We now claim that there is a space of functions on $\mathcal{O}$ which carries the G bi-module structure of End V. To an operator $T \in$ End V associate the function $\sigma_T : \mathcal{M} \to \mathbb{C}$ defined by

$$\sigma_T(v) = \langle Tv, v\rangle.$$

Since this function is independent of the phase of v, it *is* the lift of a function on $\mathcal{O}$ which we denote by a_T. Let A denote the set of all such functions on $\mathcal{O}$. It is then a fact (see [21]) that the assignment $T :\to a_T$ is $1:1$. Recall that an operator T on a complex vector space is determined by its expectation values $\langle Tv, v\rangle$ as v ranges over the unit sphere. We are stating that this is still true when the sphere is replaced by the G orbit $\mathcal{M}$.

Thus we have a canonical identification of End V with the space of functions A on $\mathcal{O}$.

Problem 3. Identify the space A geometrically.

Problem 4. How does the G bi-module structure of A, which it inherits from the fact that End V is a G bi-module, manifest itself in terms of the geometry of the action of G on $\mathcal{O}$?

A related question, which ties in with the theory of $*$ -products (see [3], [5]) is

Problem 5. How does one describe the algebra structure of A, also inherited from End V as above, in terms of the geometry of $\mathcal{O}$?

Until these questions have been answered, the 'construction' of the G bi-module A is admittedly abstract, but it is of some interest perhaps that we need only functions on the coadjoint orbit, not sections of a line bundle, to exhibit the space. For the case of $G = SU(2)$ and the irreducible representation of dimension n we may be more concrete. The space A turns out to be all spherical harmonics of degree up to and including $n-1$ on the sphere of radius $n-1$; a space of dimension n^2.

Problem 6. Develop an analogous theory for the discrete series of a non-compact semisimple Lie group.

5. Construction of Representations

The construction of irreducible unitary representations of a Lie group G is an important problem in non-commutative harmonic analysis. For compact and nilpotent Lie groups, this problem is treated as solved in the literature. It is considered unsolved, for example, for noncompact semisimple groups.

But let us stop for a moment and ask– What does it actually *mean* to 'construct a representation'? This is a meta-mathematical question, at least to the extent that one rarely finds a proper definition of the term 'construct' in the literature.

To clarify the discussion, consider once again the case of a compact simple Lie group G. Weyl determined the irreducible characters of such a group in the 1930's. He was certainly aware of the following 'method' of constructing a representation from a character.

(1) Determine the space W of functions on G spanned by all left and

right translates of the given character– this is a G bi-module.
(2) Decompose W when viewed as a G left-module into irreducibles– we get dim V copies of a single irreducible V.
(3) Take any one of the constituents—this is the required left-module.

Is this a valid construction? Well, no, since the credit for the construction of the representations goes to the Borel-Weil theorem, which only appeared some decades later. But this is a sociological answer, not a mathematical one. Mathematically, one would argue that the instructions (1), (2) and (3) are not specific enough. To what extent is the space W constructed by declaring it to be the span of all left and right translates of a character? How exactly does one decompose this bi-module into irreducibles? Can one exhibit in an explicit fashion some non-zero vector in the final representation space? A basis?

These seem quite reasonable objections, but cannot similar queries be raised about the concreteness of geometric quantization? How does one explicitly construct a line bundle over some integral coadjoint orbit? And what does this mean? How does one determine the holomorphic sections of such a bundle? Can one exhibit in an explicit fashion some non-zero vector of the final representation space? A basis?

There is some scope for disagreement and controversy here. Let us therefore try to define more precisely what we might mean by the metamathematical term 'to construct a representation.'

I tentatively propose the following:

Definition 9. *A (finite-dimensional) representation of a group G is* constructed *if a computer program can be exhibited which will (1) input group elements (in whatever form the group has been 'given')*
(2) output matrices which represent those group elements in some arbitrary but fixed basis of the representation space.

Of course it is reasonable to require only that explicit instructions for creating such a program be given, not the program itself, at least if we can agree that the instructions are indeed explicit enough. (One should also give some thought to how one 'represents' the real numbers which might appear as the matrix entries). Similarly we will say that one has constructed all the representations of a group when one has a (larger) program which in addition to the above, also inputs the representation (by some label such as the highest weight, or the integral coadjoint orbit $\mathcal{O}$). One has constructed the representations of all simple compact Lie groups if one has a (yet larger) program that in addition to the above, inputs the group G.

I realize that this proposal will not give universal pleasure. Are there any sensible alternatives? I suspect that some physicists view our abstract constructions of representations as not completely the full story–witness the large number of papers in physics journals concerned with describing explicit bases of representation spaces. It would be unfortunate if young mathematicians miss out on a chance to contribute to this area of physics because of an imprecision of terminology.

If we accept the above tentative definition, some interesting problems present themselves. Primary among them is the following.

Problem 7. Construct the irreducible unitary representations of a compact simple Lie group G.

Here are some others.

Problem 8. Given an irreducible representation of G on V, describe the geometry of the orbits of G on V.

Problem 9. How does one do *trigonometry* on such orbits?

Problem 10. Describe the hypergroup structure of the G orbits on V.

References

[1] D. Arnal and J. Ludwig, *La convexité de l'application moment d'un groupe de Lie*, J. Funct. Anal. 105 (1992), 256–300.

[2] Z. Arad and H. Blau, *On table algebras and applications to products of characters*, J. of Alg. 138 (1991), 186–194.

[3] F. Bayen, C. Flato, C. Fronsdal, A. Lichnerowicz, and D. Sternheimer, *Deformation Theory and Quantization I. Deformations of symplectic structures*, Annals of Physics 111 (1978), 61–110.

[4] W. R. Bloom and H. Heyer, *Harmonic Analysis of Probability Measures on Hypergroups*, De Gruyter, Berlin, (1995).

[5] M. Cahen, S. Gutt and J. Rawnsley, *Quantization of Kähler manifolds. I. Geometric interpretation of Berezin's quantization*, J. Geom. Phys. 7 (1990), no.1, 45–62.

[6] A. H. Dooley and N. J. Wildberger, *Harmonic analysis and the global explonential map for compact Lie groups*, (Russian) Funktsional. Anal. i Prilozhen. 27 (1993), no. 1, 25–32 tranlsation in Functional Anal. Appl. 27 (1993), no. 1, 21–27.

[7] C. F. Dunkl, *The measure algebra of a locally compact hypergroup*, Trans. Amer. Math. Soc. 179 (1973), 331–348.

[8] G. Frobenius, *Über Gruppencharaktere*, Gesammelte Abhandlungen, Vol. III, Springer-Verlag, Berlin, 1–37.

[9] V. Guillemin and S. Sternberg, Symplectic techniques in physics, Cambridge University Press, 1984, Cambridge.

[10] N. E. Hurt, Geometric Quantization in Action, Mathematics and its Applications Vol. 8, Reidel, 1983, Dordrecht.

[11] R. I. Jewett, *Spaces with an abstract convolution of measures*, Adv. Math. 18 (1975), 1–101.

[12] A. A. Kirillov, *Elements of the Theory of Representations*, Grundlehren der math. Wissenschaften 220, Springer-Verlag, Berlia, 1976.

[13] G. Mackey, *Harmonic analysis as exploitation of Symmetry*, Bull. Amer. Math. Soc. 3 number 1 (July, 1980), 543–698.

[14] J. Sniatycki, Geometric Quantization and Quantum Mechanics, 1980, Springer-Verlag, New York.

[15] R. Spector, *Mesures invariantes sur les hypergroupes*, Trans. Amer. Math. Soc. 239 (1978), 147–165.

[16] V. S. Sunder and N. J. Wildberger, *On discrete hypergroups and their actions on sets*, preprint (1996).

[17] R. Thompson, *Author vs Referee: A case history for middle level mathematicians*, Amer. Math. Monthly 90 No. 10 (1983), 661–668.

[18] M. Vergne, *A Plancherel formula without group representations*, in Operator Algebras and Group Representations Vol. II, Neptune, (1980), 217–226.

[19] N. J. Wildberger, *Hypergroups and Harmonic Analysis*, Centre Math. Anal. (ANU) 29 (1992), 238–253.

[20] N. J. Wildberger, *Finite commutative hypergroups and applications from group theory to conformal field theory*, Contemp. Math. 183 (1995), 413–434.

[21] N. J. Wildberger, *On the Fourier transform of a compact semisimple Lie group*, J. Austral. Math. Soc. (Series A) 56 (1994), 64–116.

[22] N. J. Wildberger, *Convexity and representations of nilpotent Lie groups*, Invent. Math. 98 (1989), 281–292.

[23] N. J. Wildberger, *The moment map of a Lie group representation*, Trans. Amer. Math. Soc. 330 (1992), 257–268.

[24] N. J. Wildberger, *Hypergroups, Symmetric spaces, and wrapping maps*, in Probability Measures on Groups and related structures, Proc. Oberwolfach 1994, World Scientific, Singapore, 1995.

[25] N. M. J. Woodhouse, *Geometric Quantization*, Clarendon Press, 1992, Oxford.

School of Mathematics, University of New South Wales, Sydney, 2052, AUSTRALIA

A Limit Theorem on a Family of Infinite Joins of Hypergroups

Hansmartin Zeuner

Abstract

Let $(S_n : n \in \mathbb{N})$ be a random walk on the nonnegative integers with the convolution structure of the join of the subhypergroups $\{0, \ldots, n\}$. It is shown that (for suitable norming constants $a_n \to 0$) $a_n S_n$ converges in distribution to a nondegenerate limit if and only if the tail of P_{S_1} is a regularly, but not slowly, varying function.

1. Introduction

Limit theorems for random walks on a countably infinite discrete hypergroup have been studied by many authors (Eymard–Roynette [6], Gallardo [8], [9] and Bouhaik–Gallardo [3], [4], Mabrouki [13], Voit [14], [15], [16] and the author [17], [18], [19], [20], [21]). In all of these examples the hypergroups can be defined via sequences of orthogonal polynomials (see [12]), mostly in one variable (except in [3], [4] and [21]). In this article we study the limit behavior of random walks on hypergroups on $\mathbb{N}$ that are quite different from the above: We consider a sequence of arbitrary two-element hypergroups $J_n = \{0, n\}$ for $n \geq 1$ and define recursively $K_0 := \{0\}$, $K_{n+1} := K_n \vee J_{n+1}$ (the *hypergroup join,* a hypergroup structure on the set $K_{n+1} = \{0, \ldots, n+1\}$ defined by Jewett [10], 10.5). By construction, K_n is a subhypergroup of K_{n+1} and so the union $K := \bigcup_{n \in \mathbb{N}} K_n = \mathbb{N}$ carries a natural (commutative) convolution structure $(\mathbb{N}, *)$ having all K_n as subhypergroups.

Now let $(S_n : n \in \mathbb{N})$ be a random walk on K, i.e. S_0 is a.s. the neutral element 0 and

$$P\{S_{n+1} \in A | S_0, S_1, \ldots, S_n\} = \mu * \varepsilon_{S_n}(A) \qquad \text{for all } A \subseteq K, n \in \mathbb{N} ,$$

where μ is a fixed probability measure on K. In this article we will study the question under which conditions on a norming sequence $a_n \to 0$ the

Key words and phrases. Central limit theorem, Domain of attraction Randomized sum, Hypergroup join, Distribution of Maxima.

random variables $a_n S_n$ converge in distribution to a limit $\neq \varepsilon_0$. The possible limit laws are quite different from the case of polynomial hypergroups where standard normal, Rayleigh and Gaussian distributions occur; this stems from the fact that the infinite join K is more closely related to the max-semigroup on $\mathbb{N}$, where $\varepsilon_m * \varepsilon_n = \varepsilon_{\max(m,n)}$ for all $m \neq n$ and where the limit laws are extreme value distributions too.

We will show in the following that the possible limit distributions have densities $x \mapsto c\rho x^{-\rho-1} \exp(-cx^{-\rho})$ on $\mathbb{R}_+$ for some $c, \rho > 0$ and that the domain of attraction of this limit consists of those $\mu \in \mathcal{M}^1(\mathbb{N})$ such that $t \mapsto \mu([t,\infty[) = P\{S_1 \geq t\}$ is of regular variation at ∞ with index $-\rho$. This limit theorem is in a close relationship with the limit theorem of Fisher and Gnedenko for the distribution of maxima (see Feller [7], example (b), p. 277). It is remarkable that the possible limit laws and domains of attraction do not depend on the choice of the hypergroups J_n.

2. Notation

Let $(K, *_K)$ be a compact hypergroup with normalized Haar measure ω_K and $(J, *_J)$ a discrete hypergroup with neutral element e and involution $j \mapsto \check{j}$ (see [2] or [10] for the definitions). Then the hypergroup join is the convolution structure on the set $K \vee J := K \cup (J \setminus \{e\})$ defined by

$$\varepsilon_k * \varepsilon_j := \varepsilon_j * \varepsilon_k := \varepsilon_j \quad \text{for } k \in K, j \in J \setminus \{e\},$$
$$\varepsilon_j * \varepsilon_{\check{j}} := 1_{J\setminus\{e\}} \cdot (\varepsilon_j *_J \varepsilon_{\check{j}}) + (\varepsilon_j *_J \varepsilon_{\check{j}})\{e\} \cdot \omega_K \quad \text{for } j \in J \setminus \{e\},$$

and the original convolution of K respectively J in all other cases. This construction was introduced by Jewett ([10], 10.5), who has shown that we in fact obtain a hypergroup and also has given the following general example of this procedure ([10], example 15.1D):

(2.1). Consider a sequence $(b_n : n \in \mathbb{N})$ of real numbers with $b_o = 1$ and $0 < b_n \leq 1$ for $n \geq 1$. For every $n \geq 1$ let $J_n := \{0, n\}$ the two-element hypergroup with neutral element 0 and $\varepsilon_n * \varepsilon_n = b_n \varepsilon_o + (1 - b_n)\varepsilon_n$. Then we can define hypergroups $K_n := \{0, 1, \ldots, n\}$ recursively for every $n \in \mathbb{N}$ as the join $K_{n+1} := K_n \vee J_{n+1}$. Since K_n is a subhypergroup of K_{n+1} for every $n \in \mathbb{N}$ we obtain in a natural way a convolution on the *infinite join* $K := \mathbb{N} = \bigcup_{n=0}^{\infty} K_n$. Using the above definition of the join we see that $\varepsilon_m * \varepsilon_n = \varepsilon_{\max(m,n)}$ for $n \neq m$ and

$$\varepsilon_n * \varepsilon_n = b_n \prod_{k=1}^{n-1} \frac{b_k}{b_k+1} \varepsilon_o + \sum_{j=1}^{n-1} \frac{b_n}{b_j} \prod_{k=1}^{n-1} \frac{b_k}{b_k+1} \varepsilon_j + (1-b_n)\varepsilon_n.$$

The Haar measure ω_K on K assigns to every element $n \in K$ the mass $\omega_K\{n\} = b_n \prod_{j=1}^{n-1}(1+\frac{1}{b_j})$.

The hypergroup H_a (with $a \geq 1/2$) described by Dunkl and Ramirez [5] is the special case with $b_n = \frac{a}{1-a}$ for all $n \geq 1$ of the definition above. If $a = 1/p$ where $p \in \mathbb{N}$ is a prime number then the infinite join is related to the p-adic group (see [5], §3).

(2.2). The dual $\hat{K}$ of the infinite join (see [2], 2.2) is homeomorphic to $\mathbb{N} \cup \{\infty\}$; the characters are given by $\chi_\infty = 1_{\mathbb{N}}$ and

$$\chi_\nu(n) = \begin{cases} 1 & n \leq \nu \\ -b_{\nu+1} & n = \nu+1 \\ 0 & n \geq \nu+2 \end{cases}$$

for $n \in \mathbb{N}$. Since $\chi_\infty \chi_\nu = \chi_\nu$, $\chi_\mu \chi_\nu = \chi_{\min(\mu,\nu)}$ for $\mu \neq \nu$, and

$$\chi_\nu \chi_\nu = (1-b_{\nu+1})\chi_\nu + \sum_{\mu=\nu+1}^{\infty} \frac{1+b_{\nu+1}}{1+b_{\mu+1}} \prod_{\kappa=\nu+1}^{\mu} \frac{b_\kappa}{b_\kappa+1} \chi_\mu,$$

$\hat{K}$ is a hypergroup with respect to pointwise multiplication. The *Fourier transform* for this hypergroup is defined for every probability measure P on $\mathbb{N}$ as the function on $\hat{K} = \mathbb{N} \cup \{\infty\}$ with

$$\mathcal{F}P(\nu) := \int \chi_\nu \, dP = P(\{0,\ldots,\nu\}) - b_{\nu+1}P(\{\nu\}).$$

(2.3). We will now consider a random walk on the infinite join K with jump distribution $\mu := P_{S_1}$. The distribution of S_n is then $P_{S_n} = \mu^{*n} := \mu * \cdots * \mu$ (n factors). Since K_m is a subhypergroup of K for every $m \in \mathbb{N}$, it follows from the Kawada–Ito theorem for compact hypergroups (Bloom, Heyer [1]) that the distributions of the generalized sums S_n converge to the normalized Haar measure ω_{K_m} if $P\{S_1 > m\} = 0$ and $P\{S_1 = m\} > 0$. This measure satisfies $\omega_{K_m}\{k\} = b_k \prod_{j=k}^{m} \frac{b_j}{1+b_j}$ for $k \geq 1$ and $\omega_{K_m}\{0\} = \prod_{j=1}^{m} \frac{b_j}{1+b_j}$. We are therefore only interested in the case where μ is not supported by finitely many points of $\mathbb{N}$.

3. Limiting distributions and domains of attraction

(3.1) Theorem. *Let $S_0, S_1, S_2, \ldots$ be a random walk on $K = \mathbb{N}$ such that the function $x \mapsto P\{S_1 \geq x\}$ is of regular variation in the sense of Karamata [11], but not of slow variation.*

Then there exist a sequence of positive real numbers $a_n > 0$ and $\rho > 0$ such that the sequence $a_n \cdot S_n$ converges to a distribution on $\mathbb{R}_+$ with Lebesgue density $x \mapsto \rho \cdot x^{-\rho-1} \exp(-x^{-\rho})$.

Proof. Since $x \mapsto P\{S_1 \geq x\}$ is a decreasing function, it varies regularly with exponent $-\rho < 0$ (see Feller [7], pp. 275–277). Choose s_n in such a way that $n \cdot P\{S_1 \geq s_n\} \to 1$ as $n \to \infty$. Then it follows from the definition of regular variation (Feller [7], (VIII.8.5)) that $\lim_{n\to\infty} n \cdot P\{S_1 \geq xs_n\} = x^{-\rho}$ for all $x > 0$. This implies $\lim_{n\to\infty} n \cdot P\{S_1 \geq xs_n + j\} = x^{-\rho}$ for all $j \in \mathbb{N}$ since for every $\varepsilon > 0$ we have $xs_n + j \leq (x+\varepsilon)s_n$, if n is big enough. From this we obtain $n \cdot P\{S_1 = \lceil xs_n \rceil + j\} \to 0$ as $n \to \infty$.

It follows from (2.2) that

$$\mathcal{F}P_{S_1}(\lceil xs_n \rceil + j - 1) = 1 - P\{S_1 \geq xs_n + j\} - b_{\lceil xs_n \rceil + j} \cdot P\{S_1 = \lceil xs_n \rceil + j\}.$$

Hence for $j \in \mathbb{N}$ and $x > 0$ we obtain

$$\lim_{n\to\infty} \mathcal{F}P_{S_n}(\lceil xs_n \rceil + j - 1) = \lim_{n\to\infty} (\mathcal{F}P_{S_1}(\lceil xs_n \rceil + j - 1))^n = \exp(-x^{-\rho}).$$

For every $m \in \mathbb{N}$ it follows by inspection of the definition of the characters that

$$1_{[0,m]} = \sum_{j=0}^{\infty} \frac{1}{b_{m+j+1}} \prod_{i=0}^{j-1} \frac{b_{m+i+1}}{1 + b_{m+i+1}} \cdot \chi_{m+j}.$$

Since the product is at most 2^{-j}, the convergence is uniform. Furthermore this speed of convergence implies that for every $x > 0$

$$
\begin{aligned}
P\{S_n/s_n < x\} &= P\{S_n \leq \lceil xs_n \rceil - 1\} \\
&= \int 1_{[0,\lceil xs_n\rceil-1]}\, dP_{S_n} \\
&= \int \sum_{j=0}^{\infty} \frac{1}{b_{\lceil xs_n\rceil+j}} \prod_{i=0}^{j-1} \frac{b_{\lceil xs_n\rceil+i}}{1+b_{\lceil xs_n\rceil+i}} \cdot \chi_{\lceil xs_n\rceil+j-1}\, dP_{S_n} \\
&= \sum_{j=0}^{\infty} \frac{1}{b_{\lceil xs_n\rceil+j}} \prod_{i=0}^{j-1} \frac{b_{\lceil xs_n\rceil+i}}{1+b_{\lceil xs_n\rceil+i}} \cdot \int \chi_{\lceil xs_n\rceil+j-1}\, dP_{S_n} \\
&= \sum_{j=0}^{\infty} \frac{1}{b_{\lceil xs_n\rceil+j}} \prod_{i=0}^{j-1} \frac{b_{\lceil xs_n\rceil+i}}{1+b_{\lceil xs_n\rceil+i}} \cdot \mathcal{F}P_{S_1}(\lceil xs_n\rceil + j - 1)^n \\
&\longrightarrow \exp(-x^{-\rho}).
\end{aligned}
$$

Since pointwise convergence of the distribution functions implies weak convergence, this completes the proof of the theorem if we set $a_n := 1/s_n$. ■

We conclude the discussion of this limit theorem on the infinite join hypergroup by showing that the distributions described in the theorem and the remark above are the only possible limit distributions on this structure and that the domain of attraction consists exactly of the distributions described in the theorem.

(3.2) Theorem. *Suppose that $P\{S_1 < a\} < 1$ for all $a > 0$ and that there exists a sequence of real numbers $s_n > 0$ such that S_n/s_n converges in distribution to a non degenerate probability measure on $\mathbb{R}_+$.*

Then there exist $c, \rho > 0$ such that the limiting distribution has a Lebesgue density $x \mapsto c\rho \cdot x^{-\rho-1} \cdot \exp(-cx^{-\rho})$ and the function $x \mapsto P\{S_1 \geq x\}$ is of regular, but not of slow variation.

Proof. We show first that the sequence $(s_n : n \in \mathbb{N})$ converges to infinity. For, if this is not the case, there exists a subsequence $(n_k : k \in \mathbb{N})$ such that s_{n_k} converges and hence $P_{S_{n_k}}$ converges toward a measure $Q \in \mathcal{M}^1(\mathbb{N})$. This implies that $\mathcal{F}P_{S_1}(\nu) \to \mathcal{F}Q(\nu)$ and therefore $\mathcal{F}Q(\nu) \in \{-1, 0, 1\}$ for all $\nu \in \mathbb{N} \cup \{\infty\}$. Since $\mathcal{F}Q$ is continuous at the unit character χ_∞ we have $\mathcal{F}Q(\nu) = Q(\{0, \ldots, \nu\}) - b_{\nu+1}Q(\{\nu\}) = 1 = \mathcal{F}Q(\infty)$ for some $\nu \in \mathbb{N}$. This implies $Q(\{0, \ldots, \nu\}) = 1$ and therefore $P\{S_1 \leq \nu + 1\} = 1$, in contradiction to the assumption.

Let F be the distribution function of the limiting distribution of S_n/s_n and let $x > 0$ be a point of continuity of F. Then we have

$$P\{S_n = \lceil xs_n \rceil + 1\} \leq P\{S_n < (x+\varepsilon)s_n\} - P\{S_n < xs_n\}$$

if $n \geq n_o(\varepsilon)$. This implies $\lim_{n\to\infty} P\{S_n = \lceil xs_n\rceil + 1\} = 0$ and therefore

$$\begin{aligned}\lim_{n\to\infty} (\mathcal{F}P_{S_1}(\lceil xs_n \rceil))^n &= \lim_{n\to\infty} \mathcal{F}P_{S_n}(\lceil xs_n\rceil) = \lim_{n\to\infty} P\{S_n < \lceil xs_n\rceil + 1\} \\ &= \lim_{n\to\infty} P\{S_n/s_n < x\} = F(x).\end{aligned}$$

This implies $\lim_{n\to\infty} n \cdot (1 - \mathcal{F}P_{S_1}(\lceil xs_n\rceil)) = -\ln F(x)$ for all $x > 0$ where F is continuous. Since F is nondegenerate, it follows from Lemma 3 on page 277 in Feller [7], that $x \mapsto 1 - \mathcal{F}P_{S_1}(\lceil x\rceil)$ is of regular variation and there exist $c > 0$ and $\rho \geq 0$ such that $F(x) = \exp(-cx^{-\rho})$. It is clear that also $\lim_{n\to\infty} n(1-\mathcal{F}P_{S_1}(\lceil xs_n\rceil + j - 1)) = F(x)$ for every $j \in \mathbb{N}, x > 0$.

By using the same decomposition of $1_{[0,m]}$ in terms of the χ_{m+j} as in the proof of Theorem (3.1) we obtain that $\lim_{n\to\infty} n \cdot P\{S_1 \geq xs_n\} = \exp(-cx^{-\rho})$ and hence $x \mapsto P\{S_1 \geq xs_n\}$ is of regular variation. Since F is a distribution function, $\rho = 0$ is not possible, and the tail probability is not of slow variation. ∎

References

[1] W. Bloom, H. Heyer, *Convergence of convolution products of probability measures on hypergroups*, Rend. Math (3), Ser. VII **2** (1982), 547–563.

[2] W. Bloom, H. Heyer, *Harmonic analysis of probability measures on hypergroups*, de Gruyter, Berlin, New York, 1995.

[3] M. Bouhaik, L. Gallardo, *Une loi des grands nombres et un théorème limite central pour les chaâines de Markov sur $\mathbb{N}^2$ associées aux polynômes discaux*, C. R. Acad. Sci. Paris **310, Sér. I** (1990), 739–744.

[4] M. Bouhaik, L. Gallardo, *Un théorème limite central dans un hypergroupe bidimensionnel*, Ann. Inst. Poincaré **28,1** (1992), 47–61.

[5] C. F. Dunkl, D. E. Ramirez, *A family of countable compact P_*-hypergroups*, Trans. AMS **202** (1975), 339–356.

[6] P. Eymard, B. Roynette, *Marches aléatoires sur le dual de $SU(2)$*, Analyse harmonique sur les groupes de Lie. Proceedings 1973-1975. ed. by P. Eymard Lecture Notes in Mathematics vol. 497, Springer Verlag, Berlin, Heidelberg, New York, Tokyo, 1975, pp. 108–152.

[7] W. Feller, *An Introduction to Probability Theory and Its Applications, Volume II*, John Wiley & Sons, Inc., New York, Chichester, Brisbane, Toronto, Singapore, 2nd edition 1971.

[8] L. Gallardo, *exemples d'hypergroupes transients*, Probability Measures on Groups VIII, ed. by H. Heyer, Lecture Notes in Mathematics vol. 1210, Springer Verlag, Berlin, Heidelberg, New York, Tokyo, 1984, pp. 68–76.

[9] L. Gallardo, *Comportement asymptotique des marches aléatoires associées aux polynômes de Gegenbauer et applications*, Adv. Appl. Prob. **16** (1984), 293–323.

[10] R. I. Jewett, *Spaces with an abstract convolution of measures*, Adv. Math. **18** (1975), 1–101.

[11] J. Karamata, *Sur un mode de croissance regulière*, Mathematica (Cluj) **4** (1930), 38–53.

[12] R. Lasser, *Orthogonal polynomials and hypergroups*, Rend. Mat. Ser. VII **3** (1983), 185–209.

[13] M. Mabrouki, *Principe d'invariance pour les marches aléatoires associées aux polynômes de Gegenbauer et applications*, C.R. Acad. Sc. Paris, Série I (19) **299** (1984), 991-994.

[14] M. Voit, *Laws of large numbers for polynomial hypergroups and some applications*, J. Theor. Prob. **3, No. 2** (1990), 245–266.

[15] M. Voit, *Central limit theorems for a class of polynomial hypergroups*, Adv. Appl. Prob. **22** (1990), 68–87.

[16] M. Voit, *Central limit theorems for random walks on $\mathbb{N}_o$ that are associated with orthogonal polynomials*, J. Multiv. Anal. **34** (1990), 290–322.

[17] Hm. Zeuner, *Laws of large numbers for hypergroups on $\mathbb{R}_+$*, Math. Ann. **283** (1989), 657–678.

[18] Hm. Zeuner, *The central limit theorem for Chébli–Trimèche hypergroups*, J. Theoretical Probab. **2** (1989), 51–63.

[19] Hm. Zeuner, *Moment functions and laws of large numbers on hypergroups*, Math. Z. **211** (1992), 369–407.

[20] Hm. Zeuner, *Invariance principles for random walks on hypergroups on $\mathbb{R}_+$ and $\mathbb{N}$*, J. Th. Probab. **7** (1994), 225–245.

[21] Hm. Zeuner, *Limit theorems for polynomial hypergroups in several variables*, Probability measures on groups and related structures XI, ed. by H. Heyer, World Scientific, Singapore, 1995, pp. 426–436.

Email: zeuner@informatik.mu-luebeck.de, hmz@math.uni-dortmund.de; Institut für Mathematik der Medizinischen Universität zu Lübeck, Wallstraße 40, D–23560 Lübeck, Federal Republic of Germany

MIX
Papier aus verantwortungsvollen Quellen
Paper from responsible sources
FSC® C105338

If you have any concerns about our products,
you can contact us on
ProductSafety@springernature.com

In case Publisher is established outside the EU,
the EU authorized representative is:
Springer Nature Customer Service Center GmbH
Europaplatz 3, 69115 Heidelberg, Germany

Printed by Libri Plureos GmbH
in Hamburg, Germany